INVISIBLE LOGIC

隐形逻辑

张为平 著

*Hong Kong, as
Asian Culture of Congestion*
——香港，亚洲式拥挤文化的典型

东南大学出版社
·南 京·

图书在版编目（CIP）数据

隐形逻辑：香港，亚洲式拥挤文化的典型/张为平著.—南京：东南大学出版社，2009.4（2014.9重印）
ISBN 978-7-5641-1619-4

Ⅰ.隐… Ⅱ.张… Ⅲ.城市空间—空间规划—研究—香港 Ⅳ.TU984.11

中国版本图书馆CIP数据核字（2009）第040272号

隐形逻辑——香港，亚洲式拥挤文化的典型
张为平 著

责任编辑	张 煦
文字编辑	莫凌燕
稿件统筹	杜曼萍
责任印制	张文礼
装帧设计	张为平　顾晓阳　余晓莉

出版发行	东南大学出版社			
地　　址	南京市玄武区四牌楼2号　　邮编 210096			
书　　号	ISBN 978-7-5641-1619-4			
经　　销	江苏省新华书店			
印　　刷	江苏凤凰扬州鑫华印刷有限公司			
开　　本	960mm×652mm 1/16		版　次	2009年5月第1版
印　　张	12.75		印　次	2014年9月第4次印刷
字　　数	150千字		定　价	49.80元

（凡因印装质量问题，可直接向我社读者服务部调换。电话：025-83792328）

前 言

香港之所以不同于世界上任何其他大都会，是因为同时兼备了两个条件：超高密度和亚洲性。这是一部关于高密度状态下都市状况的研究文本，以香港为研究对象。不是常见的、关于城市的人文式个人情怀吟咏，亦非政府或者规划部门对于城市建设的官方总结，而是以开放建筑学的专业视点，对于以香港为代表的"亚洲式拥挤文化"的详细解读。

"拥挤文化"概念来自于1978年，由荷兰建筑理论家库哈斯于纽约写就的、关于曼哈顿拥挤文化的回溯式宣言《癫狂的纽约》，一本关于美国式高密度的阐释之作，以其预见性与深刻性成为理解当代都会文化的经典。曼哈顿主义至今一直持续影响着全球新城模式。而香港的高密度文化，因为亚洲的东方式传统、政体上的特异以及与其金融资本之港的本质连接，都导致其城市在实体和精神方面与曼哈顿模式呈现出迥异的局面。这也是需要对其单独进行讨论的原因所在。

超密度

荷兰被称为"欧洲的高密度城市"，那么，数十倍于荷兰密度的香港，则只有用"超密度"来进行定义了。不知于何时开始，无论是主流媒体、建筑学者或是普通大众，一旦提及"高密度"，则立刻将其与"问题"或者"都市的困境"相联系。高密度意味着：拥挤、制约、紧张、压力；高密度等同于土地的超负荷利用、资源的穷尽式开采、公共及私人空间的无止境争夺；在心理层面，高密度仅仅指向压抑与不快。加之多年以来媒体对于理想生活的设定一直是欧洲式的"阳光、空气、绿地、低密度"，对于"舒适度"的过度而一成不变的渲染，使"高密度"仿佛已经成为生存的梦魇。

然而一个一直被人们忽略却不争的事实是：在舆论界对于高密度异口同声的、清一色的声讨之下，城市化进程却悄无声息、无可阻挡地一直进行，并且愈演愈烈，呈现出加速的状态。城市化的一个集中体现即是密度的激增。过去20年内，伊斯坦布尔的人口从600万增至1300万，拉各斯从200万人口增至1600万，而中国的珠三角的人口在20年内增至3600万[1]。就连一向以低密度为傲的欧洲，近年来也在中心城市出现了密度的加速增长。预期与现实的巨大差异使我们对于这个议题的任何简单判断与结论都显得陈旧、武断而缺乏洞见。

(1) 数据来源：Rem Koolhaas. Mutations, Harvard project on the city, ACTAR, 2000

习见

人们固守了百年的对于密度的认识其实是一直沉溺于某种"思维的积习",使我们对于现实的真相视而不见。"习见"是容易堕入而不易爬出的泥沼,它将思维导向谬误而难以察觉,它成为某种障碍,使任何创意的意图都成为不可能。

香港,就淹没于各式各样的习见之中,时常被误读并被误解。因其过高密度引起的"非舒适度"遮蔽了所有其尚且隐藏的、有趣的都市性。一个易被忽略却极其重要推论是:当密度达到一种超常的状态时,城市缔造者所应对的问题亦是超常的问题,而相应的,其解决手段也必然是"非常规的"。香港的每一次建造都是各种都市力量于无意识状态中对于"都市可能"的又一次试验,它不是单一的促成,而更可能是建筑师意志、开发商野心、政府的权益和民众自发的多重因素的混合作用结果:它已经提供了丰富的经验及启示。

如果"都市化"和"密集化"成为未来不可避免的趋势,那么,我们至少应该懂得一点处理"高密度"的艺术。而香港,就是这种理论的原初提供者。

亚洲式拥挤

香港是超高密度城市的终极形式,并且,它仍然在增长。建筑高密度与人口高密度的叠加,再加上24小时全天候的运转——永不止息的喧闹。如果纽约是"拥挤文化"的原型,是高密度摩天楼现代都会的最集中范例,那么香港绝不是另一个纽约的再版,尽管在发展之初直至今日,它一直有意识地以纽约作为范本试图向其接近,却始终在各种无意识的力量作用下,使其不断偏离其原型。悖谬的是,这种接近的企图越强烈,就越无奈地发现二者之间的差异越明显。这种无意识力量,是亚洲多年积淀的本土风俗、文化积淀使然。东方传统实际上一直并未离去,它以灵魂附体的方式,一直依存于"现代性"的城市躯壳之中。

如果我们继续追问亚洲的可识别性到底意味着什么?是克里尔兄弟的古典复兴还是弗兰普顿的地域批判主义?西方学界对于中国城市的批判热点之一,一直是所谓"中国性"的缺失。但是如果深度交流则会发现无论是建筑批评家还是旅游者,对于所谓"中国性"的理解基本是一种对于传统的、图像的怀旧式臆想和主观眷恋。

实际上，无论是"现代"或"后现代"，还是全球化的"广普城市"，都无法准确描述这个城市。香港与任何其他美洲或者欧洲的大都会相比，都是完全不同的都市概念。

这已经构成了足够强烈的"东方性"——用常识难以描述，而具有自身现代化的特质。而极限条件与传统意识形态、生活方式的联姻是导致这种"特异现代性"的原因。

逻辑的都市——都市的逻辑

城市并非如通常意义上所认定的，只是客观的物质实体，它是活的。

城市可以以自己的逻辑思考、行动以及进化。它可能局部机体受损，却有自我修复功能；它可以调动各方资源，平衡各种利益，以达到自身的正常运转。尽管表面上具有随机性和不可预见性，城市却自生命开始起就不断自我完善，从不成熟走向世故。

都市的逻辑通常是隐匿的，因为所有城市中发生的事似乎都是出自人为。也正是因为这个原因，其自身的逻辑往往被忽略了。本研究将香港作为一个具有自我思索、调节和演进能力的类生命体，以全新视角揭示其掩盖于表象背后的深层逻辑。

"香港主义"

从来没有一个城市如同香港，如此悖论地将城市的多义性与建筑的贫乏性统一于一身，它是有趣与无趣的混合物。如果从纯粹建筑学的角度来检视香港的建筑，实际上大量充斥后现代的意识、商业化的审美与功能主义的拼贴，以及过度被地产所主导的设计，很少有学术意义上的建筑精品；而若论城市发展，近年来也难以有相当说服力的重大突破，例如启德机场的再利用问题荒废了10年，以及西九龙文化区的迟迟未果等等。

但是我仍然认为香港是有趣的，它的有趣之处在于其长年以来一直在极限条件下发展所衍生出的一系列应对困境的都市策略：建造上、逻辑上以及政治上。香港的都市启示存在于为匮乏的建筑理论界提供一种处理极限的、从未被认真总结和研讨过的"香港都市主义"的理论构架，而非具体的建造成果。

如今建筑学学科处于一种困境之中，一方面，建筑师依然希求依仗某种实际早已失去的对于都市的特权、一种早已大面积失效的传统设计方法进行建筑设计；或者一种对于建筑学本无力解决的社会道德的越界式承担，来继续控制城市，这种臆想与现实之间存在着巨大的差异。另一方面，在建筑理论表面上层出不穷的背后，却因为援引依据之间的彼此互借而存在着重复的关系。因而，从建筑师自身定位，至建筑理论的建立，都迫切需要一种基于现实分析的、更加鲜活却犀利的方式。

本研究对于广受诟病的高密度进行重新解读，试图通过分析处于"超高密度"状态下的香港，揭示其在多年应对各种"极限条件"的挑战下所激发出的各种非常规都市状态，以及具有启示性的、非显性的造城逻辑与智慧，并进一步探讨其于无意识状态中构建一种"极限都市主义"的可能。或许我们可以将"香港都市主义"定义为"亚洲式拥挤文化"的典型。

作　者

2008年10月

INTRODUCTION

There are two points which differ Hong Kong from any other metropolis in the world: Hyper density and Asianness. This research which takes Hong Kong as its study area, focus on the city under the high density condition, It's neither a personal emotional expression of the city, nor the official report from the government or the urban planners. It's the deep insight of the city from the open architectural view on the scheme of "Asian Congestion Culture", which is represented by Hong Kong city mode.

The concept of "Culture of Congestion" origins from Rem Koolhaas's "retroactive manifesto" for Manhattan, <Delirious New York>, in 1978, which has been the crucial text to understand the American high density. The Manhattanism which has been concluded in this book has continuously influenced the world's metropolitan mode within the last 30 years. However, the high density culture in Hong Kong influenced by the oriental tradition, the mixed regime, and its economical harbor essence reveals highly difference from the Manhattan mode. This is the reason why we need to give an independent eye on it.

Hyper Density

If we call the most dense country in Europe, the Netherlands, "high density", then for Hong Kong, which is 20 times' more than the Netherlands in density, could only be considered as "hyper density."
When ever mentioning "high-density", most people will immediately relate it to, (and only to) problems: congestion, constrains, nervous, pressure, high-density means limited land, resources, limited public and private space. It leads to nothing but tension. The mass media's rendering of the European low density life as the paradise, "sunshine, fresh air, and greenery" has pushed the prejudice further.

But the fact which nobody could deny is that: when the whole world is showing their unsatisfactory with high density, the reality becomes obscure: in every corner of the world the densification exacerbates continuously, not only the newly urbanized districts, but also including the areas which "against" densification most, like some European central cities. Again, the reality goes against people's expectation.

Manner of thinking

That's the belief people believe for hundreds of years, without any suspect. The conclusion above is mostly based on "MANNER". Normally, people used to judge the things in a certain "manner of thinking". The conclusion it leads to normally does not base on facts, but prejudice. "Manner" is a river easy to fall in, but hard to get out of; "Manner" always directs the idea to some wrong directions; "Manner" is the obstacle of any intention for creative thinking.

Every simple construction in Hong Kong is one more experiment by the co-operation with all kinds of power, challenging the extreme urban conditions. The consequence is a mixture of architects' will, public taste, profit operation and the ambition of the government. Its experience provides inspiration for new urban possibilities. If we admit "densification" of urban development is the tendency for the future, we need to discover the art of living with the density.

Oriental "Culture of Congestion"

Hong Kong is the ultimate form of "super density", what's more astonishing is that its densification still increases! If we consider Manhattan, the world's earliest skyscraper city as the prototype of "culture of congestion", then Hong Kong is obviously never a copy of New York——Even though from the very beginning of its birth, Hong Kong has always taken New York as its example, and tries to transform itself into another version of Manhattan, some unconscious power has continuously driven it far from its original intention.

The power which makes Hong Kong unique is another type of "collective unconscious". What gives Hong Kong's urbanity a strong oriental feature is the local culture, philosophy and manner, the traditional Chinese life style and ideology, even colonized by the western power, the "Asianness" has never been never really uprooted, it is always there: the oriental soul is incarnated with the modernized urban shell.

If we try to further discover what does identity mean for contemporary Asian cities? Is it close to Leon Krier's Neo Classicalism, or Frampton's Critical regionalism? The western critics always criticize the absence of "Chineseness". However, if you communicate with them further, one will find that the westerners' understanding toward "chineseness" always stay at the iconic level——their own imagination to the traditional images, the nostalgia from "other world",

The logic of the city

The city is not only with an objective entity, but alive. It could think, behave and mutate in its own logic. In spite of the apparent random, it evolves from immature to sophisticated, it balances all the powers, benefits, ideas through its self-adjusting system.

The logic of the city is normally hidden, because of all the happenings in the city seem to be done artificially. Also right for this reason, the logic of the city is normally ignored. In this research, I try to uncover the hidden logic, visualize the invisible, and find a clue for the unusual urban phenomenon.

"Hong Kong-ism"

There is no other city in history could be like Hong Kong, which combines so paradoxical urban extremes: the maximum multiple urban layers with the incredible architectural mediocracy. It is a mixture of abundance and monotony.

Under the extreme condition, Hong Kong has already generated the urbanism of "how to deal with the obstacles in high density"; the real value of Hong Kong's city exists in the potential to provide a possible urban theory, rather than the actual construction results. The only problem is: this urbanism is still anonymous.

The architectural theory is in a embarrassed situation: On one hand, the architects still dream to rely on the traditional concept of design, (such as permanence, axis, dogma) which have already been proved ineffective in the reality to solve the urban problems, or they still indulge in taking some moral standards as their first concern but which actually has nothing to do with architects' own profession. On the other hand, under the apparent richness and quantity of newly born architectural theories, there is a crisis of the repeating quote of the clichés: The theorists are not able to find new vocabulary for architecture.

This research is supposed to uncover the still hidden logic from the city itself, to give some fresh blood to the already pale body of architectural theory. It is a urbanism based on reality. Maybe we could name it as "Hong Kong-ism".

目 录

前 言 1

1. 消费社会中的自我彰显 1
      1.1 平面标识 4
      1.2 楼梯标识 13
      1.3 品牌标识 23

2. 高密度生存：杂交与共生 35
      2.1 杂交 37
      2.2 共生 49

3. 垂直都市主义 57

4. 暧昧不明的公共空间 79

5. 非正式：被忽略却无处不在 97

6. 效率最大化 115

7. 边界状态 139

8. 附文：密度相关三则 165

后 记 186

Contents

Introduction 5

1. Self identification in the Consuming society 1
   1.1 plane signage 4
   1.2 stair as guiding signage 13
   1.3 branding signage 23

2. high density living:hybrid and coexist 35
   2.1 hybridization 37
   2.2 coexist 49

3. sheer urbanism 57

4. ambiguity in the public realm 79

5. Informal:ignored but pervasive 97

6. maximum efficiency 115

7. border condtion 139

8. appendix:three essays about density 165

The Epilog 186

1

消费社会中的自我彰显

self identification in the consuming society

标识的功能是将自我与他人区分，将自我信息广而告之于路人，将潜在的目标消费者引向他们应该去的地方。在一个以商业作为支柱产业的社会中，标识成为所有形式的商业的重要的自我指认符码，在一个地价与密度都如此之高的城市中，标识是生存的工具：
1. 为了指引，它必须具有空间上的连续性；
2. 为了被识别，它必须特异，同时又具有某种特定的风格和形式；
3. 为了吸引眼球，它必须鲜明、清晰，甚至还需要一点点幽默。

The function of the sign is to identify and differ yourself from the others, to advertise your information to the passers-by, to lead the related people to the only place where they should go.
In a metropolis with the commercial industry as its mainstay, the signage become extremely important self-logos for all forms of store; In a city with such a high density and land value, the signage are the instruments to survive.
In order to guide, they must be arranged spatially with continuity;
in order to identify, they need to be unique and with certain fixed style and format;
In order to be attractive, they demand impressiveness, clarity and even, a bit humor.

标识系统：基本规则

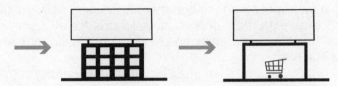

标识的基本功能：将目标客户吸引入建筑中进行商业交易。
Basic function of the signs attached to the building normally direct the customers to the commerce in the building.

标识的附加功能：某些标识将人群引入"不易被发现的区域"。
The signs on some buildings guide the customers to the store exsiting in some "not easy to be found" places.

知名品牌往往不需要大型标识，商品本身即是最好的广告。
The famous brands normally do not need sign boards, the commodity itself is the best advertisement to attract the customers.

标识尺度与建筑类型
Proportion of the signs to the buildings

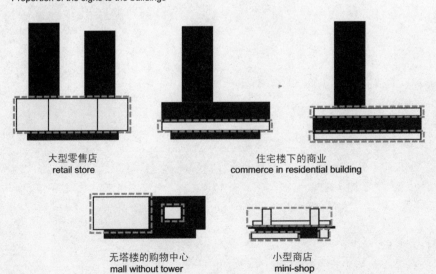

大型零售店
retail store

住宅楼下的商业
commerce in residential building

无塔楼的购物中心
mall without tower

小型商店
mini-shop

1.1 平面标识

plane signage

plane signage 01
街角广告平台放大（铜锣湾）

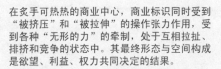

在炙手可热的商业中心，商业标识同时受到"被挤压"和"被拉伸"的操作张力作用，受到各种"无形的力"的牵制，处于互相拉扯、排挤和竞争的状态中。其最终形态与空间构成是欲望、利益、权力共同决定的结果。

多向可见性：
位于怡和街与轩尼诗道交角的一栋住宅，下部裙房被各种标识广告所包覆，可以被横向穿行的行人及快速行进的车流从不同的视角捕获。由于该点处于东西向车流和购物人流的交汇口，实数黄金地段中的黄金点，绝佳的地利条件使这栋普通的住宅裙楼成为多个商家的"必争之地。"

In the hot commercial centre, the signs are always under the tension of invisible force, they are either extruded or compressed simultaneously, their form and scale directly relate to the desire, profit and power.

Multi-orientation
Hong Kong Mansion, located at the main commercial street cross at Yee Wo St and Great George St, its podium head is fully covered by signs. They attract the pedestrian while the people go across the street. They are also visible from the vehicles passing by through the main road.

扩大展示面：
由于两条道路呈锐角相交，而这种状态显然不是理想的展示平台，因此标识牌向上、中、下三个方向伸展——过度的伸展，形成一层铺张的"悬置展示面"。品牌的知名度及商家的实力与标识的尺度、位置及角度成正比关系——展示品牌的时也展示了"做派"。

Enlarged billboard
The most brilliant brands conquer the best position of the corner: the head. The space of the corner is limited, (it shrinks to the top): obviously not an ideal platform for presentation, so the sign boards are extended both horizontally and vertically, all of which compete for more attraction. A "enlarged" advertising board is formed by over extension. The size and positive position of the signs correspond with the power of the brand.

plane signage 02
甜品店的迷你生存策略（铜锣湾）

感官的现场经营

迷你甜品店位于街角，是年轻女生和情侣最爱光顾的小店之一。以仅有的一层店面和不到20平米的营业空间，与其他大中型餐厅相比，在规模上显然处于劣势。但是它恰恰生意兴隆地怡然自得。

小店面有其独特的迷你型生存策略：全身型展示——商店立面的每个角落都被各种与其主题相关的图片、菜单、文字占满，没有任何空白区域；而商品本身的热辣与新鲜性成为最好的广告：案台上摆满颜色诱人，并且散发致命香气的即成品；另一侧堆积了各种鲜亮的水果，水滴清脆，让所有经过者难以不驻足。

所有可以吸引人的细节都被放大，以平面的和立体的、真实的和图像的各种方式，刺激路人的所有感官——一种关于"现场"的经营，以半传统、半现代的方式昭示自我——虽然犹处于无意识状态，商家实际上已经在实践一种"类格式塔"的方法：

缺乏足够的面积与尺度，惟有巧妙利用店面面格局，通过无限放大局部的作用，通过细节的累加与集中化处理来弥补自身劣势；而门前的热闹状况已经证明，效果确实不错。

Manipulating the senses

This sweet shop located in the street corner, is one of the most popular points among the youth and the lovers. With only one floor high, no more than 20m2 store space, in scale it can not compete with the large restaurants. However, it develops its own strategy to identify itself:

Every corner of the shop is covered by different kinds of images, menus, and texts. The very attractive sweet foods and fresh fruits are laid on the table with delicious smell: Who could resist its enchantment?

All the attractive details are blown up, in both plane and cubic way, both objectively and visually, stimulating all the human sense: the manipulation of perception, the merchant present its goods in a semi-tradition, semi-modern way, although unconsciously, the store owner has been practicing a quasi-Gestalt way. Lack of area and weak in scale, they have to enhance the effect of the details as much as they can. The result is not bad.

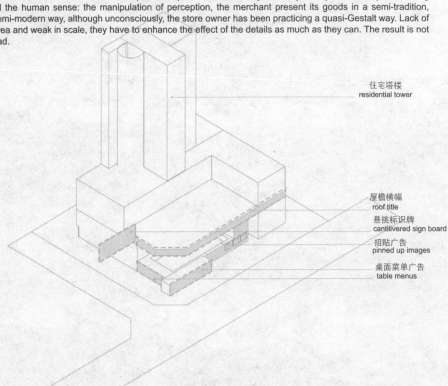

住宅塔楼
residential tower

屋檐横幅
roof title

悬挑标识牌
cantilivered sign board

招贴广告
pinned up images

桌面菜单广告
table menus

plane signage 03
广告包覆的裙房（铜锣湾）

分裂主义的二次进化

商业渗透已经势无可挡得无孔不入，只要位于或者仅仅临近主要商业街，非商业建筑也被占用作商业展示的用途。铜锣湾谢斐道民宅的底部8层裙房被各种广告标牌覆盖，而各色广告的标识并非指向建筑内部实际功能，一种新型的"分裂主义"正在诞生——建筑内部功能与表皮的二次分离。

如果说曼哈顿的摩天楼的立面的展示功能与其内部实际程式各自承担不同的作用，形成所谓"自足纪念性"，其内与外的无关联性成为分裂主义的第一层次。那么在香港，分裂主义已经进化，它更进一步：建筑不再以其立面作为展示或者象征目的，外观本身已经不再重要，退出人们视线之外，它转化为另一个更有效益的物件的载体：广告。建筑的立面退化成为纯粹的景框、背景和承载物，成为内部真实功能与外部展示之间的"中间体"和"媒介物"。建筑的立面不见天日，成为隐蔽的"第三极"。这是表与里的再分裂，建筑表面的信息可以与内部的程式毫不相关。

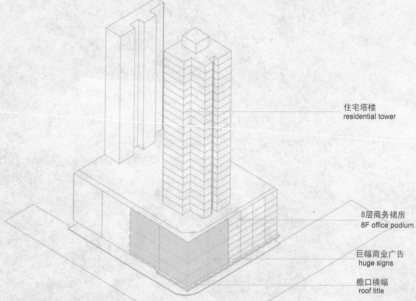

住宅塔楼
residential tower

8层商务裙房
8F office podium

巨幅商业广告
huge signs

檐口横幅
roof title

The "wrapped" podium: mutation of "Schism"

The commercial force has penetrated into every corner of Hong Kong, At the main commerce streets, even the non-commercial building will also be expropriated as the base for advertising. The potential of the 8 floor office tower's podium at the Jaffe Road cross is fully explored:

More than 70% of the walls are fully covered by all kinds of signs, no window area left. The special phenomenon here is that: Few of the signs hanging on the surface of the podium relate to the program happening inside the building. A new "Schism" is created. If we consider the "lobotomy's indispensable complement" of Manhattan skyscrapers as the first level of realization of "Schism", which has taken the exterior as purely "auto-monumentality" and independent from the internal activities, here, in Hong Kong, the Schism has already mutated: there is another layer of advertising, purely representation, outside the building facade, the facade fades away from the people's view, becoming merely the background, (or canvas); It becomes the media between the actual content and external signs.

plane signage 04
悬挑标识的森林（铜锣湾）

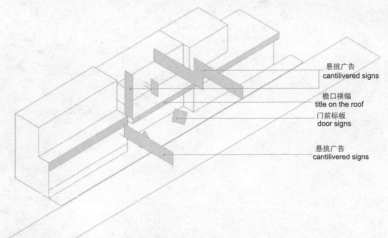

悬挑广告
cantilevered signs

檐口横幅
title on the roof

门前标板
door signs

悬挑广告
cantilevered signs

频率策略
香港沿街商业的主要特征是数目惊人的悬挑标识牌。
无数各色各类的标识"漂浮"于空中，使人有置身丛林的眩晕感。高密度、小型化、数量众多的商业体直接回应了标识的数量。每一家店面都在通过增加标识的"频率"来强调自我的"存在"。使自家店面不至于在标识海洋中被淹没——这几乎是一种面对激烈竞争刺激的应激反应。以谢斐道一间牛扒店为例，在店面不到20米的距离内，悬置了两块水平及一块竖直的标识牌"扒王之王"。相对于店面的尺度，标识的尺度不可不谓"巨大"。除此之外，店门前尚有两块门标，门头上有匾额，墙裙也附带牛扒广告。
试想商业街中每家店面都以类似的频率使用标识，那么街道上空怎可能遗留任何空隙？漫步其中，几乎有一种点彩派制造的错觉：躁动的情绪，静止的标牌似乎都在跳跃。

Forest of the Cantilevered signs: Frequency
A very special figure of the Hong Kong commercial street is the incredible amount of cantilevered boards floating in the air above the street, from which one may get the impression of the "forest of signs". Take the steak restaurant on Jaffe Rd as an example, within not more than 20m's distance (the restaurant size), there are already 2 horizontal sign cantilever and 1 vertical cantilever, besides 2 door signs, the roof title and the advertisement on the wall-skirt. Hundreds of cantilevered signs dance in the air, reaching as far as possible. Each one of them tries to be impressive, and conquers every possible orientation of people's views. It's almost an illusion created by the post impressionism painting: jumpy emotion.

1.2 楼梯标识

stair as guiding signage

stair as signage 01
楼梯=欢迎姿态（铜锣湾）

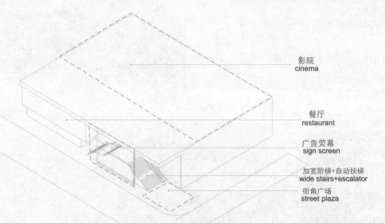

影院
cinema

餐厅
restaurant

广告荧幕
sign screen

加宽阶梯+自动扶梯
wide stairs+escalator

街角广场
street plaza

建筑的表情
建筑的布局、空间的设置能否传达情绪，或者某种态度？
在惯常的观念中，建筑等同于纯粹的"客体"和"非感性"，这一观念在此处受到质疑。
铜锣湾的一家电影院入口处，商家通过打通双层、雨篷出挑及台式楼梯的组合，制造了强烈的"欢迎氛围"。从剖面上看，楼梯与雨篷形成一个向外开放的大口，可以轻易地将目标人群"吸入"其体内（同时，也将资本大量吸入）。楼梯的宽度加宽的非常规化处理（香港的楼梯均尽可能窄以节省交通空间），踏步进行"高亮化"处理以制造奇幻氛围，这些举措进一步加强了该建筑的诱惑力。

Welcome stair: Expression
Could the architecture convey some kind of emotion?
Architecture is normally considered as non-emotional because of its pure objectivity. This view is challenged by a cinema in Causeway Bay. The stairs in front of the cinema are widely open to both sides of the street corner (in Hong Kong, the stair width is normally narrow to save space, this treatment is very unusual), together with the wide canopy and the enlarged frontage, They together show a strong gesture of "welcoming", to introduce the pedestrian to the interior cinema space. And the steps are highlighted by the lamps to be more fantastic. The architecture starts to have its own expression.

stair as signage 02
楼梯=流线控制（铜锣湾）

最大化"消化"

交通元素如楼梯、自动扶梯与通道开口方式，不仅具有传送人流的作用，在商业中它更有一层隐蔽的作用：流线控制。商家通过这些元素的设置，实现他们最大化"吸入"客流，并且使客户遍历每个角落的潜在意图。在东角购物中心，这种意图与建筑设备的结合达到极致。

中心西侧只设一部上行扶梯，将客流引入二楼寿司店。顾客用餐之后必须穿过廊道由东侧的商场电梯下楼。商家意图十分明显：用餐之后，顾客必须经过商店，增加可能的购物的几率。

而在东侧的入口门厅，表面上有上下行的开放自动扶梯，顾客可以自由进出，实际上仍然玄机暗藏——从二楼逛完下至一楼准备离去时，在出口处会迎面三个着粉裙并有白色翅膀的"兔女郎"，面带微笑的向你示意：请继续往里走——你必须逛尽一层的购物空间。

兔女郎的微笑如此动人，指引也如此优雅，使人无法拒绝她们的请求——即使这种示意的实质略带某种"强制性"：看似自由的购物空间，却处处存在"无形的控制"，对于消费者意志的过分控制，使人质疑这种诱导的正当性。一种隐形的"边界效应"：人体本身也可以成为控制的手段，而且这种手段比设备或者实体边界还有效的多。

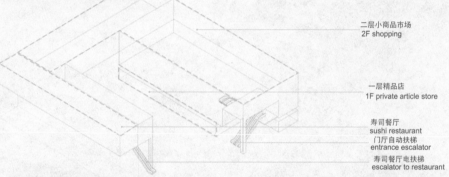

二层小商品市场
2F shopping

一层精品店
1F private article store

寿司餐厅
sushi restaurant

门厅自动扶梯
entrance escalator

寿司餐厅电扶梯
escalator to restaurant

Maximum Swallow

The main shopping space of Laforet also mainly locates on the 2nd and 3rd floor, which is above the retail brand shops. The hidden principle of the arrangement for the circulation of the customers is: "maximum swallow", and "to keep them stay as long as possible."

This is expressed in the arrangement of the escalators. The main entrance has double direction escalators bringing people up and down. But the one at the restaurant side only brings people upwards. The intention is obvious: after your dinner, you can not immediately leave, but have to go through the whole shopping space to the entrance hall. This will make more possibility for commerce. For the entrance hall, which apparently seems quite free to go in and out, the fact is also complicated. Again the merchant plays a trick with the customers: once you land at the ground floor with the escalator going down, you will come across 3 lovely girls dressing in Rabbit costumes, standing at the exit, and guide you very elegantly to go for further visit on the ground floor. It is very hard to refuse the charming smile from the rabbit girls. In order to "direct" the flow, any kind of facility could be made use of, even the human body.

折返购物流线

用餐后购物流线

兔女郎的微笑

stair as signage 03
楼梯=门面（铜锣湾）

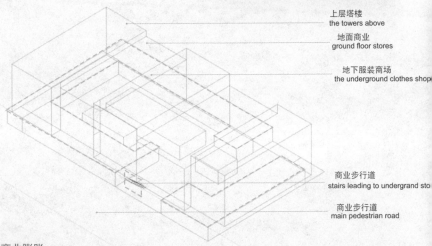

上层塔楼 / the towers above
地面商业 / ground floor stores
地下服装商场 / the underground clothes shop
商业步行道 / stairs leading to undergrand sto
商业步行道 / main pedestrian road

地下迷宫：商业膨胀

当地面上一层商业空间已被全部侵占时，只有寻求向上或向下发展大面积的商业空间，形成一种商业"膨胀作用"。

拥有相对低廉的地价、大面积可用空间和较短的距地距离，地下空间实在是除地面层之外的商业的理想选择。与其他任何城市地下层通常只作停车和仓储等功用的"灰色空间"消极处理不同，香港中心商业区的地下空间往往是活跃的购物空间。如"铜锣湾地库"——其名字已经昭示了其原有空间属性——作为地下仓储空间，如今已被改造为商场。香港地下商业最特殊之处是：在地面层，这个商场几乎是不可见的——临街仅留入口及标识：入口狭窄，标识醒目。拾级而下，顾客立刻会发现自己已经身陷一个地下商铺的"迷宫"中：完整的平面被玻璃隔断分割成一个个连续的商业单元，透明、均质，而内容则彼此互异。商店的本身即是一个展橱，承载了交易+展示的双重功能，实现了自我身份的模糊化与多义化。

很难想象，低矮昏暗的地库可以拥有如此丰富和广阔的包容性，且这种地库并非只此一家，而是处处可见。"地库商业"将城市生活加倍。

Underground labyrinth: bulging commerce

If you want to have a large shopping area, but there is no space left on the ground, what will you do? This shopping district is called "Causeway Bay underground warehouse", the name already tells you its property: it is transformed from the former underground storage space. The uniqueness of Hong Kong underground commerce is that they normally have nothing left on the ground floor except for the entrance: It's almost invisible. Entering by a very tiny stair downstairs, you will immediately land in a labyrinth like shopping space. Taking the advantage of the former storage, the flat space is further divided into a series of corridor show cases, you can hardly imagine it could be so wide and endless.

stair as signage 04
楼梯=导引（铜锣湾）

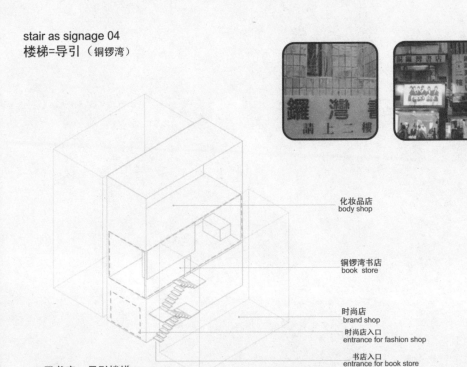

化妆品店
body shop

铜锣湾书店
book store

时尚店
brand shop

时尚店入口
entrance for fashion shop

书店入口
entrance for book store

二层书店：导引楼梯

香港商业中心区土地价值极高，首层当然是最理想的营业层，但是其高昂价格也令小型私人业主无法承受。在利益驱使下，二层或者三层通常被认为"消极"的空间也被征用，成为小型商业或事务所的栖居地。以二层商业"铜锣湾书店"为例，不到20平米的面积，完全重叠于一层的时尚店之上。其首层空间被邻区挤压至"极少"状态——一个楼梯，在复杂的邻里间反复转折、破旧、并呈现出略带阴森的氛围，恍若置身于希区柯克的电影场景——它难免使人疑虑：这上面真的有书店么？

店主深知顾客心理，早有安排：从门头起始，至沿途的墙壁，楼梯，一路用明黄底色黑字的标识及书刊介绍作为引导，不断消除顾客疑虑，并以"更上一层楼"将顾客最终导入，楼梯与标识已经成为一体——商业化的楼梯必须超越其基本属性，同时具有功能性与展示性。

Second floor book store: stair=guide

In the central shopping area, the land value is extremely high. The ground floor is doubtless the most attractive and ideal space for the commerce, but it's also relatively hard for the private merchant to gain. Encouraged by the profit, the floors above are also made use of as shopping space, tiny but efficient.

The book store is not more than 20m2, it fully overlaps on the fashion-shop below. While the fashion shop has conquered the full width of the single building size, the owner of the book store has to make a independent stair way leading to its own shop. The stair is narrow and horribly mysterious, making people hesitate to entre; but by the assistant of varieties of "signs", they inspire you finally step into the lifted store.

书店流线

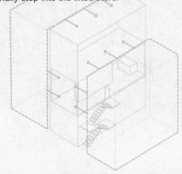

邻里挤压

1.3 品牌标识
branding signage

brand signage 01
LV主题 （铜锣湾+尖沙嘴+中环）

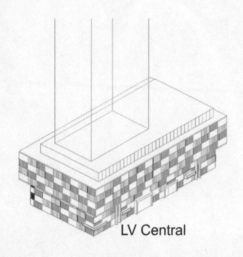

LV Central

LV Tsim Sha Tsui

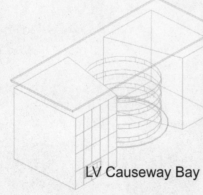

LV Causeway Bay

均奏与变奏——重复中的差异

作为消费文化的核心，品牌在宣扬自我的战略上必须处理一个永远的悖论：如何既能保持各个店面的连续可识别性又能宣扬其个性？

Louis Vuitton在香港主要商业区：中环、铜锣湾及尖沙嘴都有"据点"。品牌店自身的发展要求其各个店面具有连续性，类似的主题，反复强调其品牌特性，而时尚的精神却是"永远求新求变"，这两者的平衡即是时尚店考量的焦点。

方格是LV的基本主题，三地的店面都将其作为基本的"网格"，而填充方式和材料却完全不同。

中环店用不同反射角度的金属细条构成，因不同的观望角度和光线条件产生精致的细分效果；尖沙嘴店则用实体的石材包覆，以不同凹凸程度产生光影变化；而铜锣湾店则用颗粒纹玻璃填充方格，配以淡紫色灯光营造优雅氛围。

对于时尚店，表皮与建筑的关系同衣服与人体的关系几乎可作类比——以尽可能优雅个性的方式，将建筑包覆，如同择衣：应同时满足自身的愉悦与他人的审美认同。

Repetition and Difference

One brand will always insist on its own "style" for its store design. To enhance the "identity", they need to keep certain coherence.

However, the essence of "fashion" is to be creative, fresh, and "forever new." If it only repeats one standard, the brand will be dead. So how to produce difference, at the same time keep certain recognizable feature is the main focus of the store designers.

Louis Vuitton uses "rectangle" as its basic scheme, but differs from each other in different occasions. In Central shop It can be vertical metal strips with reflective angles, In Tsim Sha Tsui it becomes solid panels with different depth; In Causeway Bay shop it changes to little purple glass dots filled in a rectangular grid.

brand signage 02
Fendi的装饰性构造（尖沙嘴+中环）

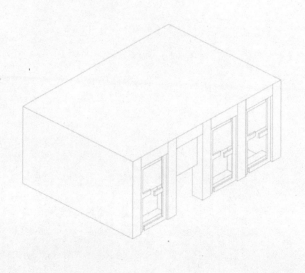

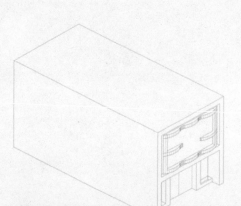

商标=构造=装饰
一个品牌设计的智慧在于对Logo的灵活运用。Fendi利用其标识的完形特性，尝试将其与店面的构造结合，同时具有装饰效果。Fendi的标识由"F"与倒置的"F"构成——巧合地（或是有意而为的），它同时是一个稳固结构：作为立面的柱的一部分。当商标成为结构，它的"放大"也有了足够的理由——这同时也是对其可识别性的放大。这种结合式操作是极巧妙的：它呈现出一种半装饰半构造、半平面半立体、半本体半展示的双重效果。
它是消费主义的么？它是表皮么？似乎都是又似乎都不尽然。商业设计的本质本来就是一种"诡辩"的设计，它是标识的展示性与构造的工具性的深度结合。

Logo=Construction=Ornament
The logo of Fendi is combined of an "F" and an inversed "F". It's a stable structure, when the logo becomes part of the construction, the store owner naturally has an excuse to enlarge it. This transformation is a wonder, in the sense that it becomes a mixture of advertisement, structure and decoration. It's plane but also cubic; it's ontological but also representative; Is it architectonical or "post-modern"? Is it purely commercial or also in some way artistic? Seems all fit but none of them could cover the whole. The store design is indeed some sort of dialectical design, in this case, it's the combination of representation value of logo and the instrumentality of construction.

brand signage 03
Dior的氛围渲染（尖沙嘴）

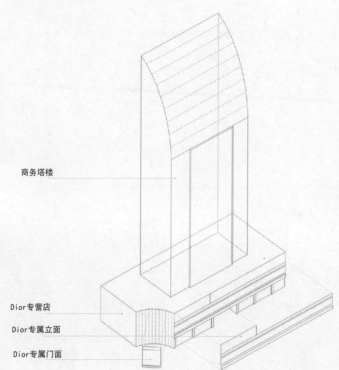

商务塔楼

Dior专营店
Dior专属立面
Dior专属门面

完全占据
名品店的强大氛围渲染力表现在以下三个方面的执着：
1　全局一体化：标识以各种方式在店面、店身的各个可能的地方重复——经过含蓄而精致的设计，往往隐含于特定的标识中，半透明、时隐时现。字体与色彩本身已经具有高度的审美趣味，与展示的对象成为整体。
2　从任意方向接近，你都不可能错过。虽然只占有裙房部分，可是Dior以其强大的品牌"渲染力"将整栋楼，甚至其周边都纳入其自我的范围之内。半独立的部分与半结合的整体之间的张力是对于城市的一种激活。在这个意义上，Dior尖沙嘴店成为一种城市的"独立的部分"，与其他一般城市肌理相区分；而一切又以如此谦逊的方式进行：一种精心制造的"不着意感"。
3　"拒绝式的灵韵"：一种只有名品店独有的现象是：门口有一位穿制服的营销人员把守，控制人流量，顾客必须排队才能进入。全世界的商家都是"打开门来做生意"，只有这里是采取了一种近乎"拒绝"的态度。实际上，这正是名品店的高明之处：刻意营造所谓的"高档"氛围，一种难以亲近的、人造的"高贵感"——面向高端消费者，它有意提示自身作为"艺术"的价值而非"普通商品"。

Fully occupied
For one brand, the creation of "brand atmosphere" is important.
Even in a high-rise, the brand is able to cover the whole building through its radiant effect, even it occupies only the podium part. Take Dior Tsim Sha Tsui store for instance: Firstly, the brand logo is repeated many times in different forms in all positions: the title, the gate way, the beam, the column, the frame...No matter from any direction you access to the building, you can not miss the information: " Dior is here." Secondly, its own pattern is hidden in the facade and space design itself. They do it in such a subtle way that one could hardly feel the pattern, they have already melt and penetrate into every corner of the store. Thirdly, the "Sedulous Elitism": One phenomenon which you could only found in the well-known brand flag-ship store is that the amount of customers is controlled by the sales-lady at the entrance, you will have to wait in a queue. This intentional "artificial elegance" immediately increases the attraction of the brand.

brand signage 04
CK 摩天广告楼（中环）

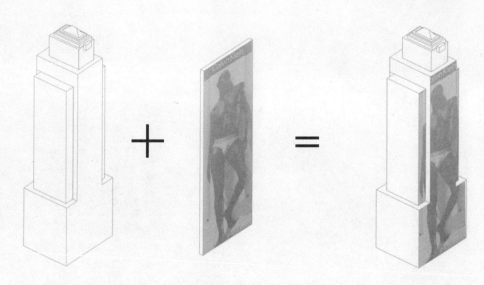

街道标识的极大版：对于都市视觉景观的独占
Calvin Klein征用了一整栋摩天楼作为其广告载体，位于中环的皇后像广场：一个裸体的男星，拥有完美身材和黝黑肌肤，只着一条ck内裤挺立于广场一侧。这栋高层高达40层，成为超尺度的广告牌。选择尖塔型高层作为底板并非偶然，高耸的塔楼就是对于"性"的隐晦象征，以强调ck带来的性感。这种提示是隐而不宣的，而这种模糊性正是绝妙的宣扬效果。
在消费主义浪潮以不可阻挡之势席卷一切的态势下，名品似乎也可以不须任何掩饰地张扬自我；于全球资本时代。品牌成为了城市景观最具统治性的因素，审美、意识、生活方式……一切皆受其主导。一种被商家设计的意识形态，全部的人群已经无可避免的卷入。从广告的角度评价，这个摩天广告无疑是有效而有趣的，可是这种对于都市视觉景观的霸权式占用和滥用，是否是我们这个时代文化缺失的一种警示？

Skyscraper Billboard
CK takes the whole facade of a skyscraper as its billboard.
The super scale advertisement is more than 100m high, with a naked black star wearing only CK underwear. The choice of minaret like skyscraper is not accidental. The stately and imposing aloft tower is a perfect symbol of masculine feature. The relationship is ambiguous. The ambiguity is right what CK demands.
In the age of global capitalism, the "brand" becomes the most dominant power to decide the cityscape. The aesthetics, ideology and life manner......nothing can escape from this tendency, which changes our life so deeply. As an advertising concept, the sky-advertising board is interesting, however, the over expropriation of the city view by the commercial signs also reveals a crisis of cultural supremacy.

brand signage 05
欲望的缩影：便利店（湾仔+铜锣湾+筲箕湾）

都市基本欲求的相似性

便利店使人惊讶之处在于：虽然在港便利店的品牌众多，隶属于完全不同的商家，可是却在尺度、室内陈设及商品种类上呈现出高度的一致性。

为人群提供日常居家用品，或者街头逗留时的"不时之需"，以支持都市生活的运转。如果说便利店的经营者必须足够敏感，而且货品种类一定基于多年营业经验总结，那么如此惊人一致的结果，恰恰就是当代都市人群生活欲求的精确反映。货品种类可以分为应对公共生活（街头）和私密生活（居家）两类。普通日常用品为：饮料、香烟、零食及报刊杂志；而应对突发状况的热门物品有药品，安全套及雨伞。在尺度相近的空间中，所不同的只是货架的角度，和收银台的方位。

便利店如同现代都市生活的一个"百宝箱"，可以随时打开、随时取。仔细检视，我们发现，原来都市生活的欲求竟然如此简单而无差别。

Convenience shop: similarity of urban desire

The amazing thing of the convenience shop is that no matter which company or brand they belong to, the layout of the store and the type of commodity, reveals highly similarity and coherence.

They provide the most popular goods people need when they are on the street, or something urgent. The common daily used things are: drinks, tobacco, snacks and magazines. For emergency there are: medicine, condom, umbrellas. In front of the store, there are normally a shelf of newspaper and a row of little gambling machine. The things in the convenience shop show the basic requirement of the modern life: easily consumable, pragmatic, and the inpermanence.

2

高密度生存：杂交与共生

high density living: hybrid and coexist

密度作为一个热点话题，被讨论已久，并且拥有大量相关成果。似乎再对其目录进行任何新的补充都已显多余；似乎任何新的讨论都已被讨论过，并且过时：人们对于这个主题已经过于熟悉而产生厌倦。它似乎是一个已终结的主题，一个不再热的热点。
——但是它仍然是一个从未有结论的话题。
土地的承载力究竟有多大？如何在地块内容纳尽可能多的居民，又兼顾日照、通风和舒适度？
欧洲居民对于高密度无法容忍，他们不能将其与舒适和便利连接，他们坚信，密度越低越舒适——这已经是一种成见。
高密度并不一定指向消极因素，至少，它提供的便利性，高效和资源的集中化，生活的丰富层次，在低密度状态下是无法获得的。

Density has become a hot issue, under all kinds of discussion for long, with a lot of reports about. It seems to be out of fashion to give any new comments to this catalogue, anything you discuss seem to be discussed and out of time, people are bored of the topic. It is a terminated topic, a hot spot never hot again.
However it is also a topic which never leads to a conclusion.
What is the capacity of the land? How to accommodate maximum amount of people, with amenities and enough daylight, ventilation?
Once coming across with high density, the European people have disbelief and can never relate it to comfort and convenience. In their definition, the lower the density is, the more comfort you get. This is already a prejudice. High density is not necessarily equal to negative effect, it is not a so urgent "problem" which need to be solved immediately, on the contrary, it provides many advantages the people live in low density areas will never be able to enjoy.

2.1 杂交

hybridization

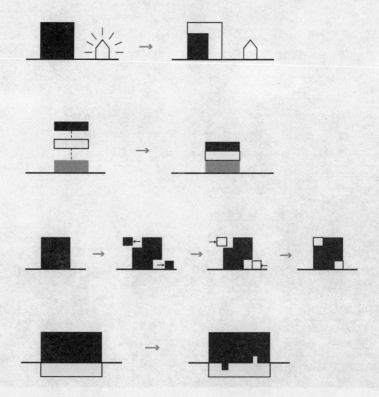

hybridization 01
堆积式社区（筲箕湾）

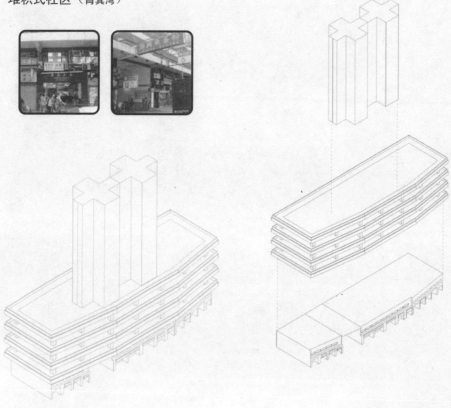

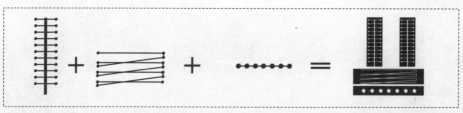

堆积式杂交
惯常的空间思维方式在超高密度状态下不再有效，建筑程式以非常规的的方式组合叠加——基于效率与利益优先原则。事物的组合方式被进行变更，达到料想不到的特殊结果——虽然往往起始于被动，却终结于意外的收获。
以荃湾的一栋住宅为例，它的特异之处在于：整个"小区"的概念被集中实现于单栋建筑中。三种现代住区的主要功能：居住、商业服务、停车被全部叠加于地面之上。在叠加功能的同时，它也同时叠加了三种完全不同的空间组织模式：联立式、螺旋式及垂直树状结构，各自保留独立入口。通常置于地下的停车空间，在此处有多达6层的需求，置于地下显然不经济，因此被抬升至地面；而底层的商业价值又使其不能存在于一层，于是"夹心停车场"成了唯一选择——堆积背后所隐藏的逻辑。

Stacking as hybrid
Stacking, which means the programs are mixed in a unusual way, is based on the principle of efficiency and benefit. The normal logic loses its effect in the high density condition, one of the typical housing block in Tsuen Wan, its uniqueness is represented by the condensing of the community in one building block: the residence, the parking and the commercial are all above the ground, stacking on top of each other. It is also the superimposition of 3 types of spatial mode: juxtaposition, rhizomatic, and spiral.

hybridization 02
侵染式学校住区（西环）

受环境左右的杂交
在高密度状态下，建筑的属性不再是一成不变的；受到环境的不同程度影响，建筑的程式或属性可以随时间而变化。阿尔多.罗西的类型学在高密度状态下失效，建筑不再具有清晰的类型——即使最初的指向明确，也可能随着时间推移而变成混杂，难以言明；且这种变动是一种动态过程，永远处于一种"临时状态"，一旦有新的决定力量介入，即可继续进行进一步的变更。在这个西环的住宅群房内的教会学校的案例中，高层住宅的裙房部分竟然是一座教会学校，而学校得以存在，影响显然来自近邻住宅一侧的教堂；作为精神承载体的建筑，不论体积多小，一旦拥有宗教的力量，总是具有强大的辐射能力。

"Infected" hybridization
The apparently "nothing to do with each other" programs: school and housing, by the pressure of limited land, can be combined together by super imposing. The different programs run independently among the vertical spatial separation. The school is the media which communicates between the church and the housing. It also indicates its own "schism": Spiritually, it corresponds to the religious building; Functionally, it serves for education for the residence in the community. Once given the spiritual power, no matter how small it is in scale, the building will immediately gain the radiant effect (the communal church).

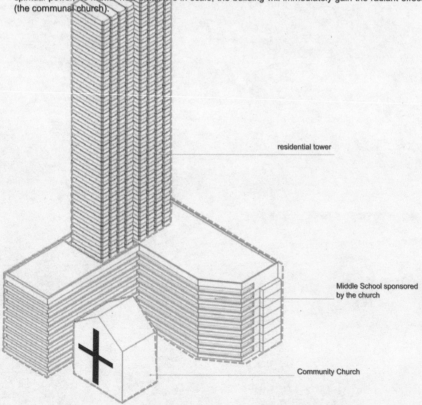

residential tower

Middle School sponsored by the church

Community Church

hybridization 03
演化式城市标本（湾仔）

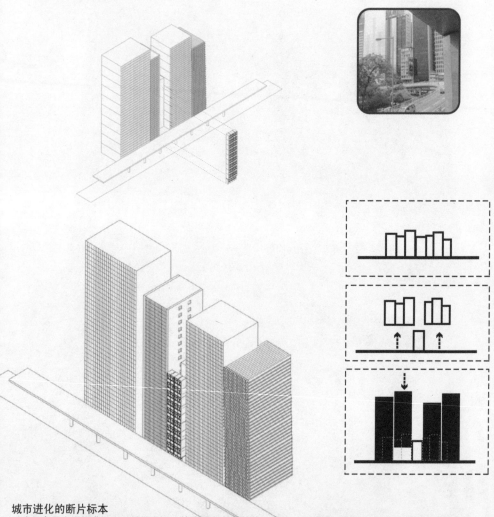

城市进化的断片标本

沿湾仔东区走廊行进，可以赫然发现一栋浅绿色的窄楼栋"贴附"于其后的众多办公建筑的玻璃幕墙之上。它的类型、尺度与破旧程度都使其如同一个天外来物，与其后的办公楼的华丽和规整形成强烈对比、格格不入——然而它就在那里。在多年前，该区都是与其同类型的住宅，其后随着城市发展，旧住宅被成片移除，大型摩天办公楼取而代之。而这栋住宅因为某些特殊原因得以保存，于是形成奇观式的都市景象。
城市演化的残留物，是新与旧的并置、巨大与狭窄的并置、华丽与破旧的并置。城市改造及更新的印记，如同一个凝固的标本，将过去与现在并置于同一时空中。这里的拼贴现象并非柯林.罗的"拼贴城市"，而是亚洲都市普遍存在的、近乎"超现实"拼贴化现象的一个范例。

Sample of urban evolution

Driving along the lifted highway parallel with the coast of Causeway Bay, you will suddenly get shocked by a view: Among the forest of sumptuous business skyscrapers, there is a 9 floor, old fashioned residential building standing right in front. With cheap construction, monotonous unit arrangement and slim volume, this housing has very sharp contrast with its surroundings. It seems absurd, eccentric. But, it is right there.
It's almost "surreal". This caricature alike image shows the paradox during the urban development: the new office building starts to eat the land, however with some certain reason, this little piece of housing was able to survive during this battle. It now coexists with the office towers in an almost "attaching" way. This is an example of "time based hybridization", it could be seen as a helpless choice, but it also gives inspiration of how to "exist with the others".

hybridization 04
叠加式：城市层（中环）

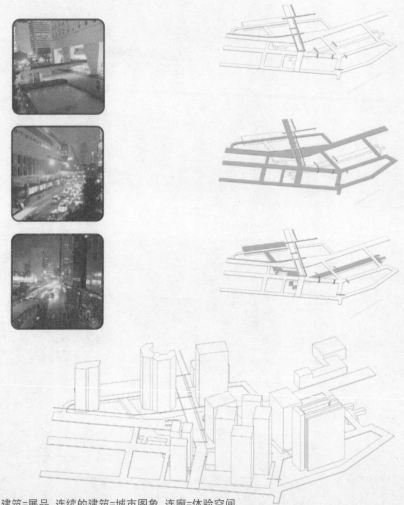

中心=秀场　建筑=展品　连续的建筑=城市图象　连廊=体验空间
中环的空间组织实际上是一场精心策划的"都市秀"，秀的受众主要是香港的初到者及游客，唯一目的是：使你震惊。看似无心其实有意，这条路线起始于中环码头以及香港站——所有交通工具的真正枢纽（机场快线、地铁、码头），无论是从水路或是地铁，一旦走出该站，初到者立刻身陷于一片惊人的繁华都市的奇观组合中：向北可以越过海滩眺望尖沙嘴的天际线；往南则由连续的廊道及天桥将你带入高楼大厦的丛林中，金碧辉煌的立面、喷泉、花园，以及脚下永无止息的车流；一切与"繁华"有关的要素都可以于此找到。世人对于都会景观的所有期许与梦臆，如今已经呈现在你面前。廊道如同一个将各种断片连接的蒙太奇工具，精心设计的路径，主导观看的顺序，情感也随着所见达到高潮。一个长时间的，过度兴奋的旅程之后，这段历程终结于一个完美的性感句点：兰桂坊。这是利用建筑作为布景的佳例，成功地使外来者在步入香港的第一时间，立刻为其"都会氛围"折服。

Urban show: Central
The whole organization of the "central" area is actually an "Urban Show" for the Hong Kong new comers, the only aim is to "make you shocked": The linear path starts from Hong Kong station, once you step onto the island, you immediately get yourself lost in the amazingly prosperous urban images, to north you get the open view of Tsim Sha Tsui, to the south the continuously running overpass bring you through the skyscrapers' forests, with fountains, gardens, with vehicles running as never-stop river. The dream of this metropolitan one ever has is right in front your eyes. And after a over exciting trip, the journey terminates in a perfect sexy ending--Lan Kuai Fong.

hybridization 06
嵌入式：地铁名品综合体（中环）

基建与商业的联姻
这是交通设施与商业的机智结合，中环的地铁出口之一恰恰就在LV旗舰店的首层。由于其中只有一个出口，并且与整个店面设计合一，它几乎没有影响店面效果，相反，却将人流直接引入置地广场。
这个结合带来的隐蔽性，以一种优雅的方式，掩盖了地铁站的突兀，同时掩盖了商家的野心。商业与交通的联姻，体现了消费社会的都市权利来自于资本：城市的交通干线连接的两极是住区和商场。

Insertion: metro exit & brand store
The Exit of the metro is right in the flagship store in Central, The commercial makes the visiting route go through their own spaces--A perfect combination of infrastructure and commerce. They make use of the long distance of exit path, to create maximum meeting points for all the stores. The metro-brand shop operates with a double gain strategy: it hides both the ugliness of the metro entrance and the ambition of the merchant, a marriage of commerce and transportation.

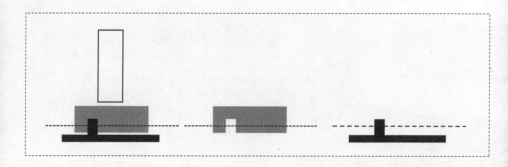

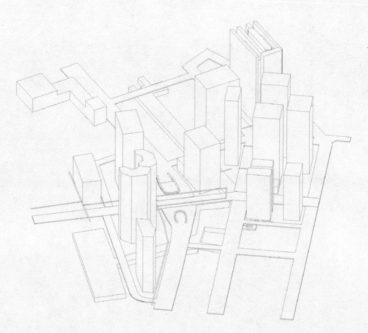

2.2 共生

coexist

Coexist 01
并置式&与跃层式（兰桂坊）

联立式
一种最常见的并存类型，各个参与个体以平行相接方式依次并置——以相邻的个体来看，中间物通常处于被"包夹"状态。在兰桂坊的实例中，一般为单层，体量相似，坐落于裙房底层。个体上，它们是彼此互异的情节载体；而作为群体，它们是相似的街道"普通元素"。它们是悖论的单元：最大化个体自由与最大化群体统一性的兼容载体。

JUXTAPOSE
The frequent type: juxtaposition
The bars with one floor space, similar size, located on the podium of the block, will normally be juxtaposed to each other. Individually, they are different plots verify from each other, collectively, they are generic elements of the "street". They are the units combing with paradoxes: the maximum individual flexibility and maximum collective unity.

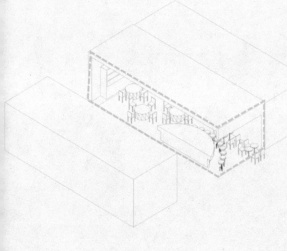

跃层式
当一层的店面因狭长而显得过于逼仄，且两侧的空间已被占满无法继续成为延展向度时，一些酒吧则向上垂直扩张。占用两层空间来补偿水平方向的拥挤，但是并不将其分割成两层营业空间，而是合并及上下贯通——利用空间错觉来形成通透的意象，获得适当的"品质"。

DUPLEX
Some of the bars occupy two floor space, in order to compensate the horizontal tightness, the floor in-between is taken off, the two floors are combined into a duplex. The horizontal tightness is compensated by the vertical enlargement.

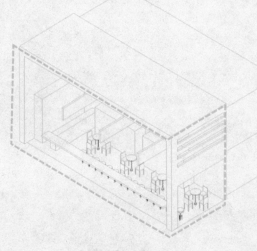

whiskey priest
an irish pub

LIVE SPORTS
ENGLISH PREMIERSHIP

FRI 26			SUN 28		
8:45pm	STOKE v MAN UTD		10pm	NEWCASTLE v LIVERPOOL
9pm	TOTTENHAM v FULHAM		10pm	FULHAM v CHELSEA
10pm	LIVERPOOL v BOLTON		10pm	ARSENAL v PORTSMOUTH
10pm	MAN CITY v HULL		12:15am	BLACKBURN v MAN CITY
10pm	ASTON VILLA v ARSENAL			

coexist 02
后院式&漂浮式（兰桂坊）

后院式

新近开发的酒吧没有机会占用沿街的店面，解决之道是在前排商店之间置入狭长的过道，在街面上留下与内部的"接口"。后院式也有优势：退居于后，往往可以获得更多的商业空间，提升了利用的自由度。这是一种"欲扬先抑"的手法，也是一种空间的"插叙"。退于隐秘的后部，空间更具神秘感与私密性。

BACKYARD

Some recent developed bars do not have the chance to have the first row by the street side, what they do is to put the very narrow corridor entrance in-between the gap between the front pubs, something good of the second row is that the developer captures relatively more floor area.

漂浮式

基于相同的逻辑，首层没有获得营业面积的商家也可以将店面以"提升"的方式置于二层——仅留入口与楼梯间在底层。店面的渗透如同利益的隐喻：它如此强大，无处不在、无孔不入。

FLOATING

Based on the same logic, the pubs which do not have chance to make business on the ground floor lift their space onto the first floor, what they leave on the ground is just an entrance and an stair case. The power of profit is so strong that it permeates everywhere.

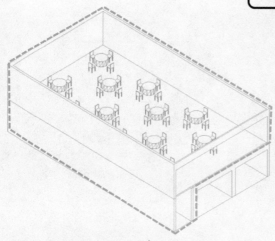

POST 9 7

WHERE PEOPLE MEET...
WITH MODERN CONTINENTAL CUISINE
AND AN EXTENSIVE WINE LIST

OPEN DAILY SERVING
BREAKFAST, LUNCH, SNACKS,
DINNER & LATE NIGHT SUPPER
9:30AM - LATE

RESERVATIONS
2810 9333

POST 97
LAN KWAI FONG,
CENTRAL, H.K.

coexist 03
角落式&坡地式（兰桂坊）

转角式
因其具有两面的可达性，转角店面的处理通常与其他类型不同：它必须两面开放；同时由于可以被人们从对面步道看到的几率也被加倍了，所以其展示与标识的指向通常也具有双重性。也许这可以被定义为"闪回"手法：以不同时间和视角经历同一件事情。

CORNER
The corner space is normally treated differently from the others. The advantage is also obvious:
It has two sides accessibility, and the chance to be viewed is also literally doubled. So this type normally opens to two directions. You could experience the same thing in different time and from different perspective.

坡地式
"坡"是港岛常见的地形，如果想获得效益，就必须学会面对"坡"，酒吧也不例外。坡道因其室内地平的倾斜而不易利用，可是商家将其作为室外酒座的平台，座椅的微小单元的灵活性足以抵消坡的倾度——将其分割为小段，坡在小范围内仍是平的；而室内则被几家店面瓜分，造成室内与室外的分离，人造地平与自然坡地的分离。

RAMP STRIP
The ramp is the normal reality of most Hong Kong land, the pubs also have to deal with it. Although the pubs are arranged with slight height difference, it does not influence the usage of them. Further more, the tables and stairs placed outside the stores play a subtle role of continuity, both visually and functionally.

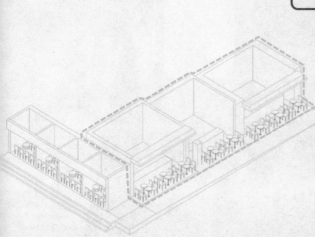

3

垂直都市主义

sheer urbanism

高密度是一种状态,但并非前提。
高密度的真正始作俑者是高地价政策。超高的地价意味着有限的土地必须容纳最大的容积率,加之香港以山地为主的自然地理条件,决定了香港的城市建设必须应对这两种超常的状态。香港建筑师已经发展了"山地摩天城市",城市景观成为摩天楼与摩天楼的叠加。

The high density is a condition, a normal status of Hong Kong. However, it's not the premise.
The real promoter of high density is the high property value policy, which is triggered by the government and controlled by the developers. On one hand, The super high land value means that limited land must accommodate maximum Floor Area Ratio, On the other hand, the hilly geographical condition is the natural obstacle confronting with Hong Kong architects.
To deal with these two extreme, usual conditions, Hong Kong architects have developed "mountain skyscraper city": Vertically the cityscape reveals the superimposing of the skyscrapers; and horizontally, it becomes the "clustering" of extremely extruded high-rises.

sheer urbanism 01
叠加的院落（香港大学）

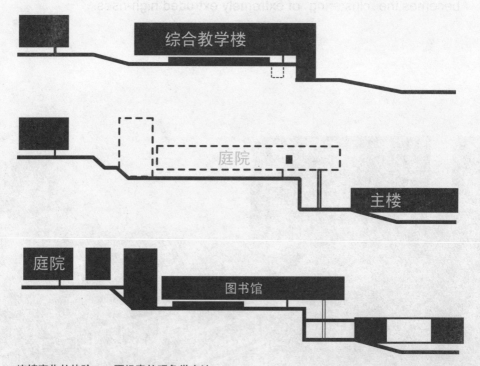

连续变化的体验——不经意的现象学方法

香港建筑师在处理地形方面是将"消极"转化为"积极"的天才——基于多年与山地周旋及在各种"不可能"地形上工作的经验，他们可以在任何坡地上建造，并且无限向上延展，创造有趣的空间条件。位于港岛西部半山之上的香港大学是一个"岛屿建造"的典型。
港大中心图书馆及周边教学楼顺坡将建筑向上及向下全局铺开布置。建筑的堆叠不仅提供了多种观望及活动平台，同时制造了平地上难以企及的、不断变化的空间体验。按照Steven Holl现象学的理论，建筑与场地的关系是"锚固"的。建筑不仅承载了其自身的功能性（教学），同时作为一种中间介质，使自然得以凸显。

The natural height difference gave the buildings of Hong Kong University a chance to be unique. Maybe the Hong Kong architects are the genius to transform the negative to positive, to find a way to make the dead alive, to make the impossible come true. Based on years' experience of struggling with all the "impossible geographical conditions", they could now build on any type of site, and extrude the building endlessly. The stepping of the building groups not only provides a large variety of terraces, and public spaces, they also give you a ever continuous changing spatial experience.

sheer urbanism 02
山地摩天楼（一）（香港大学）

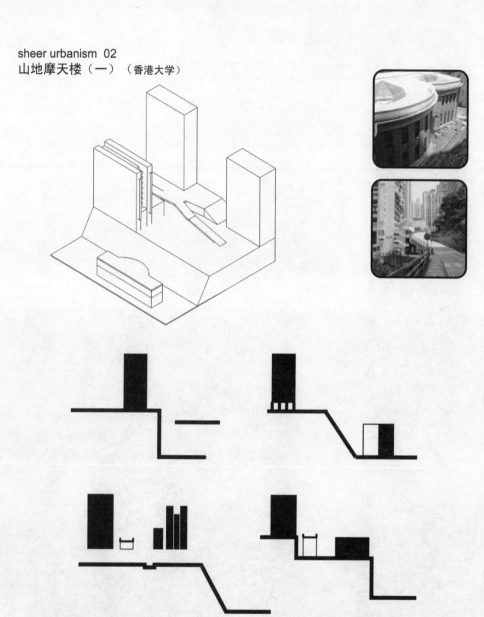

动量装置

各个校舍及各系教学楼坐落于不同高差之上，由坡道、天桥与楼梯连接。它的外在组合仿佛是为"动量"的激发而设置的——一个激发"事件"的装置。通过这些管道与媒介，学生的生活得以在各个校园的组件之间流动：通道并无特异属性，而使用者的运动使其具有意义。

在高差明显、地形复杂的区域，"连接体"的意义变得分外重要，可以根据流于其间的穿梭、停留、交错来获得校园的生活指向信息。这些入口与路径并不容易为外人了解，但是对于长期生活于此的师生来说，并不是什么问题。

Different school facilities located in different height are connected by the ramps, bridges and stairs. For the building bordered by the precipice, the accessibility becomes extremely important. The connections are not so easy to find, but for the students who study here, it's never a problem.

In the geographically complicated area, the connections have special meaning. The students are able to "flow" into every corner through these pipes. The connectors themselves are neutral, it is the movement of the users makes them special.

sheer urbanism 03
山地摩天楼（二）（香港大学）

山地按照等高线抽取并削平，用作建筑的起始与平台。右图所示的教学楼的屋顶由于其具有面向城市的极佳视野，而成为学生公共设施领地的首选，如活动中心、饭厅、咖啡店，及社团活动部都设置于此。

有了垂直都市主义，都市生活成为"永无止境"的意象——不再仅仅限于层与层之间的跨越，依托自然的地形恩赐，生活可以以整个楼栋为单位，垂直向上延续。并且离地面越远，获得的视野越丰富。

人与山地的关系、城市与自然的关系不再指向对立，山地不再意味"荒芜"，而成为"城市化的山"这至少在"可以造城的潜在地形上"已经实现了飞跃。

The roof of the lower faculty, with the open view toward the whole city, becomes the natural terrace for the campus life. There are student union, cantina, open cafe and students' associations located on it.

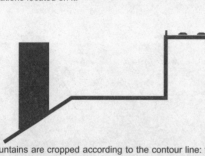

The mountains are cropped according to the contour line: to be the platform for new building. The programs are not only switched between different stories within one building, but through the unit of "a whole building"; With sheer urbanism, the city is no longer on the opposite side of "nature", the mountain becomes "urbanized mountain", at least, it already achieves a breakthrough in exploring the potential of the "buildable land".

在生化楼下巨大的混凝土桁架斜撑柱之间，竟然设有供学生休息的座椅——因其为在各个设施之间穿越的必经之路。在粗野与细腻的对立之间、在可行与不通的疑虑和坚决之间，校园的空间体验充满了变幻不定的悬疑效果——对于克服通常平地空间容易产生的单调、厌倦感是一种完全自发式的改造与冲击。

The space underneath the huge building volume also could be public. In between the scary trusses there is a platform connecting different faculties. The chairs also can be found here. The "dead corner" is transformed into some sort of human intimacy scale.

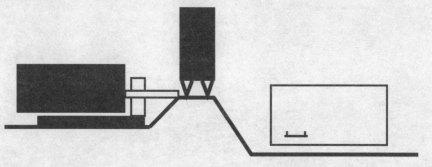

sheer urbanism 04
拉伸：超薄高层（上环）

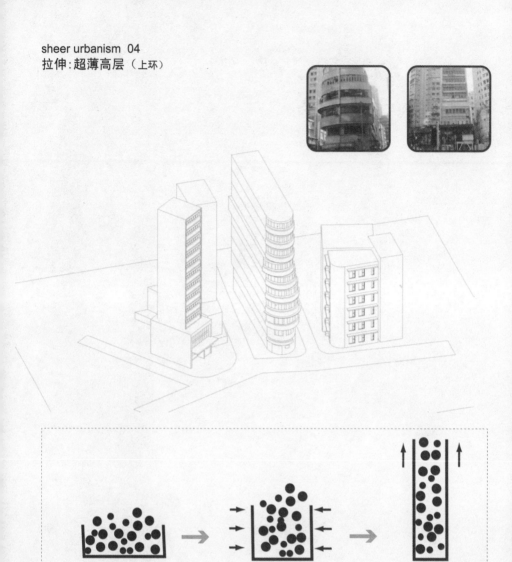

超薄——对于边界的重定义
高层建筑的最小平面单元是多少？什么才是最有效率的建造？这是香港最极端也最常见的，于有效地块无限生长的"不可言明"的秘密。仅仅以占满地块边界为基准的范围却垂直向上拉伸至极致——最小的基地面积，却有最大的程式承载力。多数被认为"无发展潜力"的地块如街角、边缘的残留地都可以被利用——每层可以只有一户或者两户，也可以共用一个垂直核向上复制20次。按照通常的建筑定义，它是"不经济"的；然而相对于原本空无一物的"边角地"，这种作为立刻显出理性与价值。这实际上不仅仅是对于建筑边界的重定义，更是对于建筑学学科范围的重新建立。

Super Thin
This is the most extreme but also most common demonstration of Hong Kong unspoken method of surviving in a limited site, simply by "extrusion" of the plot boundary to the maximum, that is, minimum plot area with maximum program capacity. The most unimaginable places, such as very tiny street corners or endings, or left over space between two buildings, could also being occupied. They are built in the unbuildable locations, in a "impossible way". The sharp edge of the buildings could be read as the boundary for different land ownerships: everyone strictly within his property.

sheer urbanism 05
悬挑：极力远伸（上环+湾仔）

悬空的潜力
在有限的地块内，当建筑已经沿地块边界无限向上复制多次，垂直方向的潜力已经被穷尽之后，还有继续增加容积率的可能么？在建筑法规允许的范围之内的唯一可能是：向外出挑——只要结构允许，可以一直出挑至法规允许的最大距离。这是对于极限的再突破，极限之上的极限。
于是香港大部分紧俏地段的沿街建筑塔楼部分都没有放过这一机会。出挑方向一致，都是朝向沿街或空旷一侧——从来不会在相邻建筑之间出现，仅仅因为已经饱和。
结果是：香港的高层纷纷以惊心动魄之势出现，是对结构承载力和大众心理承受力的双重考验。出挑包含的一个永远的悖论是：建筑一直向外探寻、占据、伸展，却从未试图加强与外的交流，只是单纯的内部利益的扩张。

The other way to gain more space is to cantilever the volume to the outside as much as you can. The ground floor is within the boundary line, following the zoning regulations, while all the floors above will be extended to add more extra FAR. This phenomenon normally happens at the street side, while there are still a bit space left. "Cantilever" never occurs in between two buildings, simply because it is already too full!. It is time to test the extreme of the structure. The permanent paradox with "cantilever" is that: the buildings always extend outwards, and conquer more space, but they never try to communicate with its surroundings: pure expanding intention for the internal benefit.

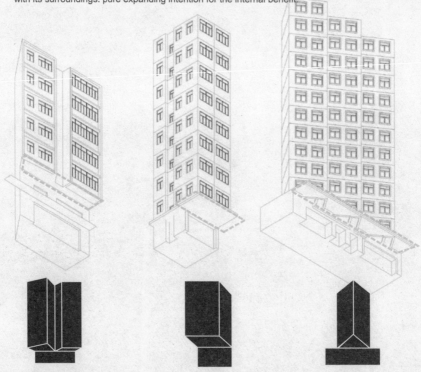

sheer urbanism 06
高密度中的空白（中环公园）

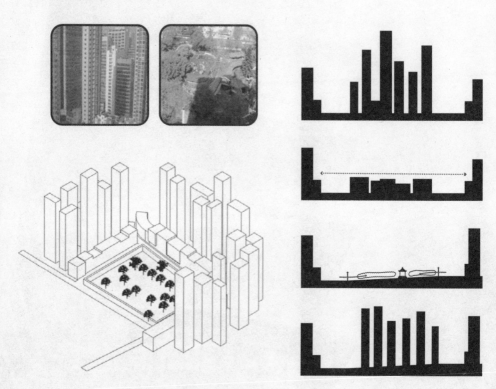

都市"空白"——高密度下的呼吸空间
在香港如此寸土寸金的地方，高密度区域内部的"都市空白"似乎是不可能的，没有存在的理由。但是事实上它们的确存在，主要有三种表现形式：1. 历史遗产、古迹。例如老中区警署，现在成为一个文化展馆，其本身也具有高度历史价值，得以存在于中环的密集高层之间。2. 宗教建筑。比如教堂和寺庙，作为市民的精神载体，其固有形式不可能改变，因此仍然得以保存。3. 城市公园（往往只是小块绿地）。
高密度状态下的"空"从来就不是空白，而是城市事件的激发器，是吸纳都市居民公共活动之水的海绵。它要么是民众自我定义、自主活动的场地，要么是灵魂的寄倚的港湾。它永远是"瞬间的"、不可重复和不断被重新定义的场所，却不断被新的程式填满，"空"是瓦解都市生活"两点一线"枯燥面的突破点。图中所示正是"中环丽柏公园"——一个试图以古典园林意象出现的社区公园。但其粗糙的设计和施工使其距离其所指相去甚远，尽管如此，市民的热情与对于公共空间的向往仍然使其充满活力。院墙将周边高层硬生生拒之门外，使内部片片土地成为高密度城市之肺。

Breathing Space
In the super high density area, the urban void seems to be unbelievable. But they do exist. There are normally 3 types of this kind of spaces which can find a place to survive:
1. The historical heritage, (such as Old Central Police Station, now as a museum);
2. Religious buildings. (such as community temples or churches) In the modern society, it seems more desirable for the people to find some place to give them a calm heart.
3. Small public spaces (such as parks). What's popular in Hong Kong is the "traditional chinese garden" style parks. The layout, the material and the atmosphere are too far from the traditional gardens, and the design is always "without design". They are nothing more than the scene set on the stage.
Nevertheless, these types of community parks are still popular among the citizens, especially the old and the children. The void within the high density condition never means "nothingness", they are the condenser of urban events, they are always "impermanent", they are refreshable and will be "redefined" and "refilled" by new programs endlessly.

sheer urbanism 07
典型平面：标准住宅（一）（将军澳）

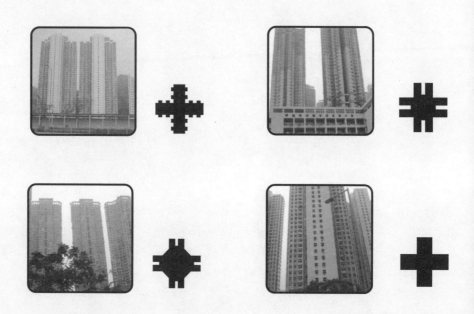

堂皇的标准化与超高
在香港，住宅由两个主要特点
1 标准化：由一种产品化的流水线生产方式生产，有惊人的"过度标准化倾向"。设计师的理由是：高地价、高容积率，要保证良好的采光和自然通风，因而"十"字形是兼顾效率与公平的最佳。
2 超高层：高地价与高密度同时扮演着"高压"角色，转而成为将住宅楼无限向上泵起的动力，无论什么户型，一般不会低于30层。这种现象在欧洲人看来，已经不仅仅是令人惊讶，它几乎等同于"非人性"、"不可居"，它被当作一种灾难、一种怪物。但是事实上，大多数香港居民都栖居于"怪物"当中，且情况并没有欧洲人想象的那么坏。相对于低密度的农会的庄式住宅，至少有一点是只有住得高才能体验的：享受高阔的视野。居住于天空中的感官感受是寄生在地表的生物难以体会的。
在这里，为什么要对被广受诟病的香港住宅提出一点不同看法？因为本研究的一个重要命题是探讨欧洲式的低密度与亚洲式的高密度究竟孰优孰劣？正因为所有对于城市的评价中，都将欧式的慵懒闲适作为"宜居城市"的唯一标准，而事实上，这种占统治地位的观点正是一种欧洲式的片面，所以这个问题才更有必要重新审视。（对于香港住宅的消极面，如逼仄的个人空间、屏风楼效应等，已经有许多研究者作过大量的探讨，在此不做赘述。）

The most frequently applied dwelling typologies share 2 common features:
1. Standard plan
This cross plan is considered as the "most economical and efficient" way to explore the full potential of the land. If you want to achieve "Far Max", and at the same time guarantee enough open view and natural ventilation for all the units, the "cross plan" is considered to be the only choice.
2.Super high rise
The high property value and the extremely high population density play the role of "high pressure", which is transformed as the "power" to extrude the buildings seldom less than 30 floors' in height.
In a European citizen's view, this is not only something "amazing", but will also even be considered as "disaster" "inhuman" and "unlivable". They are "Monsters".
However, the Monsters now accommodate most of the Hong Kong residents, and the situation seems not so bad. At least, there is one good point to live higher: you enjoy wider open views!

sheer urbanism 08
典型平面：标准住宅（二）（九龙）

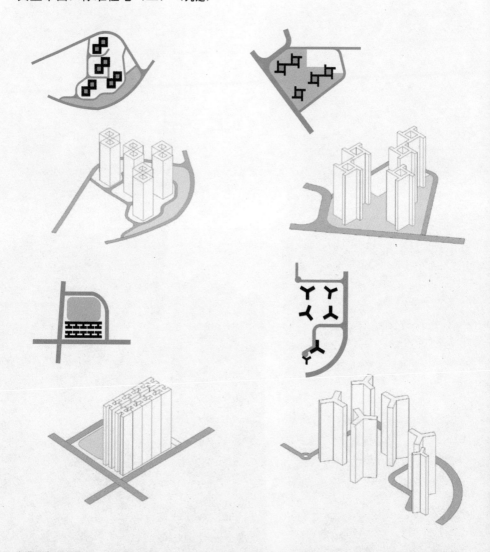

为满足高密度需求，除了最常用的"十字"平面之外，香港的住宅有几类"最受欢迎"的类型："口"、"万"、"H"及"Y"型，以适应"紧密"、"多向延伸"及"效率最大"原则。可以看出，香港建筑师及地产商在应对高密度问题的多年挣扎的过程中，已经进行了多种"类型"的尝试———一些较可行的"原型"被以结构主义的方式运用——单一元素在基地范围内重复使用，而各个"子元素"之间略有差异。遗憾的是，它具有结构主义的逻辑，却无结构主义的以人为本的核心。

Dense Dwelling Typology
To fulfill the density requirements, there are several most "popular" typologies which is widely applied in HK housing development, except for the most frequently used "cross", there are also "H", "enclosed courtyard" and "parallel". The principle is "dense", "expanding to all the possible directions" and "maximum efficiency". We could already see the effort the Hong Kong architects put in experimenting the typologies during years of struggling with high density issues. Some of the "successful" prototypes are spread repeatedly among the whole site, in a almost "structuralism" way----in the same form apparently but without the same spirit: humanism.

sheer urbanism 09
典型平面：标准住宅（三）（九龙）

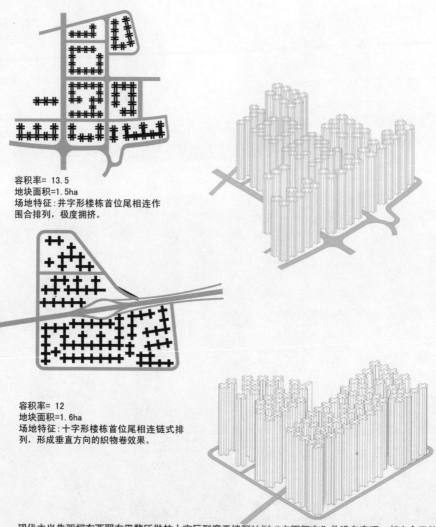

容积率= 13.5
地块面积=1.5ha
场地特征：井字形楼栋首位尾相连作围合排列，极度拥挤。

容积率= 12
地块面积=1.6ha
场地特征：十字形楼栋首位尾相连链式排列，形成垂直方向的织物卷效果。

现代主义先驱柯布西耶在巴黎所做的十字巨型摩天楼群计划"光辉都市"并没有实现，却在今天的香港备受尊重。仿佛柯布西耶的灵魂附体，十字摩天楼被以丛林式方式密植于香港的各个地块。从建筑个体角度判断，其单一的类型极其单调乏味；但若以群体视之，如此众多的同样巨塔的并立，其所具有的惊人的尺度已经具有了某种纪念性。个性的缺失伴随着集体纪念性的获取，并且将其消极面与积极面同时发挥至最大。这也再次验证了现代主义的悖论：具有一个正当的精神内核，但在实践层面却失却了人性。

Le Corbusier's proposal of "Radiant City", which is composed of Cartesian Skyscrapers, hadn't been realized in Paris, but find a new way to survive half century later, in Asia, Hong Kong, and has been widely spread everywhere. Judging from the architectural view, this is with no doubt a highly monotonous typology, however, such a large amount of huge towers standing together, they have already formed a sense of monumentality. The collective monumentality gains from the sacrificing of individual identity. Ironically, the modernism which roots in the socialist's heroism realize its real prosperity in the Capitalism society, and being enlarged to the maximum both its positive and negative aspects

sheer urbanism 10
标准住宅模式类比

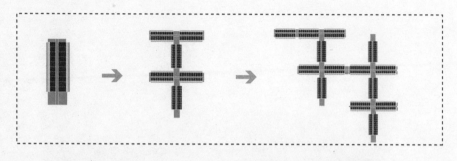

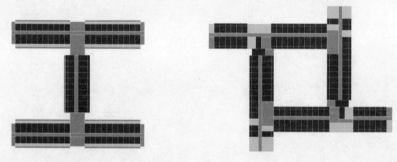

公共屋邨
常见的政府兴建的公共保障性住宅户型及组合逻辑。

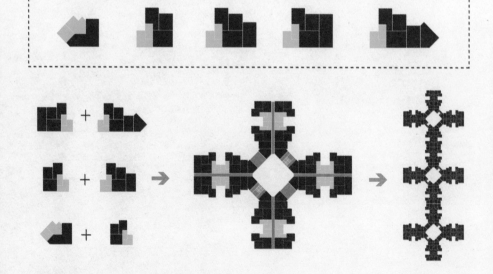

私人住宅
常见的私人住宅户型及组合逻辑。

住宅单位地域分布百分比

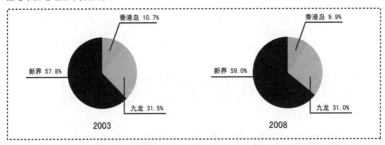

私人永久住宅价格

私人永久住宅价格	1997 港元/平方米	2002 港元/平方米	2007 港元/平方米
香港岛	78581	32777	58915
九龙	61605	24372	44284
新界	59685	24556	33693

按房屋类别划分的人口分布

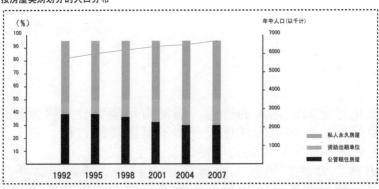

房屋租户平均居住面积

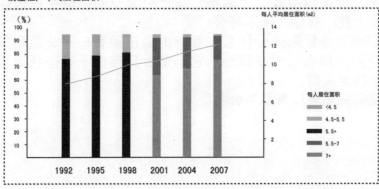

数据来源：香港政府统计年报

4

暧昧不明的公共空间

ambiguity in the public realm

在商业化程度如此之高的香港，仍然有可能为公众保留真正的"公共空间"么？如果全球化是一种不可逆转的过程，那么这种潮流是否最终只能导致由利益驱使的公共空间？
它们仍然"公共"的么？或者只是一种由"不可见之手"操控的、另一种形式的造钱机器？在香港，这些问题的答案似乎不那么简单，而是很微妙。"消费"与"公共空间"似乎已经变得你中有我，我中有你，密不可分。
公共空间的终极形式是什么？香港的确还拥有一些非盈利目的的公共空间，那么它们在何处？它们是否只存在于某些特定时间，服务于特定人群？
公共空间究竟可以有多"公共"？

In a highly commercialized society, is it still possible to maintain some space for the public?
If globalization is a un-inversed process, does this tendency necessarily increase the crisis of mono-profit oriented public spaces? Are they still public? Are they just money making containers operated by the "hidden hands?"
Couldn't the public space be free any more?
In Hong Kong, the answer seems not so simple, it is subtle. The relationship with public space and consume becomes not dividable. They are the body and the shadow, and they are solved in a dynamic balance.
If "shopping" is the ultimate form of public activity, What is the ultimate form of public space?
Hong Kong does have some other form of public realms, non-profit. Where are they? Do they only belong and serve for certain people in certain time?
How "public" can public space be?

ambiguity in public realm 01
时代广场：公共vs私利

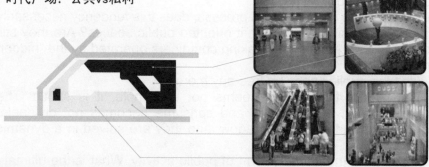

2008年初，一场商家与民众关于"时代广场"地面层公共性的争议使香港的"公共空间"问题浮出水面。其实"公共与商业"的关系一直是香港长久以来难以清晰定义的问题。根据最初的协议业主获得土地开发及税制优惠的前提是：建成后必须保证商场底层以及店前广场作为公共空间完全免费向公众开放。最初商家也在相当程度上实现了承诺。而随着时间的推移，业主渐渐"暗渡陈仓"将部分底层空间出租给星巴克之类的盈利性空间，并且门前广场的台阶及花坛也有保安拒绝行人憩坐，这就引发了公众对于其"公共性"的质疑。在公众的压力下，如今情况有所改善。纯粹公共利益的获得在商业社会中是否尚有可能？是否在当代资本社会中，公共性已经不可避免地必须与商业成为"双生"的连体儿？我们如何在与商家的博弈中争取这种与商业利益相关的公共性？它是消费社会中一种可行的做法么？

At the beginning of year 2008, there is a protest about the using of ground floor of Time Square for private use. According to the contract between the developer and the government, the ground floor should only be used for public purpose, then the developer could get some preferential policy such as tax discount.
But recently the client started to make it for private profit and commercial exhibition.
This is a paradoxical issue: Is that possible to gain purely public benefit? Is it possible to get a balance between these two things?

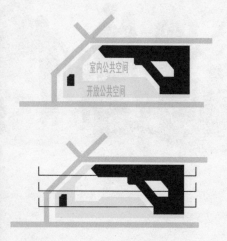

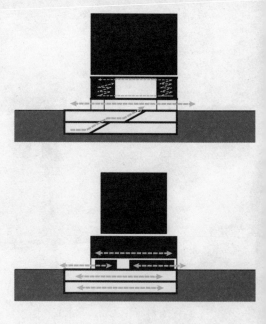

ambiguity in public realm 02
公众热情：社区公园（一）（西湾河）

在西湾河沿主街的老街区的衰败背后，掩藏着沿海铺陈的住宅新区。高架的道路成了天然的边界：越过边界，便从前一个二十年转入了下一个二十年。住宅依然是常见的"典型平面"——以高地价为完美托辞的最大化十字平面，它被称为"兼顾了效益、视线、自然通风的唯一选择。"——一个建筑师为自己的创造力匮乏制造的动听借口。然而这并不是此处讨论的重点。

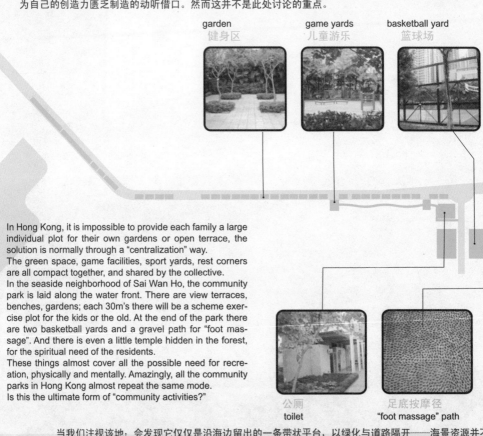

garden 健身区
game yards 儿童游乐
basketball yard 篮球场
公厕 toilet
足底按摩径 "foot massage" path

In Hong Kong, it is impossible to provide each family a large individual plot for their own gardens or open terrace, the solution is normally through a "centralization" way.
The green space, game facilities, sport yards, rest corners are all compact together, and shared by the collective.
In the seaside neighborhood of Sai Wan Ho, the community park is laid along the water front. There are view terraces, benches, gardens; each 30m's there will be a scheme exercise plot for the kids or the old. At the end of the park there are two basketball yards and a gravel path for "foot massage". And there is even a little temple hidden in the forest, for the spiritual need of the residents.
These things almost cover all the possible need for recreation, physically and mentally. Amazingly, all the community parks in Hong Kong almost repeat the same mode.
Is this the ultimate form of "community activities?"

当我们注视该地：会发现它仅仅是沿海边留出的一条带状平台，以绿化与道路隔开——海景资源并不"专属"于住区，实际上已经被交还与大众。越过这片绿带，你就进入另一个世界，于其中，无尽的自由得以存在。它已经为居民及游人提供了在香港中心区极少见的户外生活的资源。亚洲住区一直以"安全性"为名义，默许某种私有化的专属圈地，住区被樊篱围绕并有保安把守，非住区公众是不得其门而入的——尤其是所谓的"豪宅"区。这无疑造成了福轲所论述的空间的"权力等级"。而在西湾河，住宅的价格与档次也可谓不菲，却比较人性地实行了居住与休闲（景观带）的分治：封闭式的居住+开放式景观的模式，没有将景观也"围入"其私有圈地，是对于市民公共性的充分理解。介于私有与公共之间，是香港公共空间暧昧性的另一种体现。

The Sea

ambiguity in public realm 03
公众热情：社区公园（二）　（西湾河）

滨海道路系统地联结所有的景观节点，及活动设施——保证无论从哪个设施观望，都可以获得对于海的直接体验——仅仅单纯地沿道路漫步，仍可以保持开阔的心情。有趣的是，在英文中，"地块"同时具有"情节"（plot）的含义。那么，我们是否可以把地块内所发生的定义为"情节"？在西湾河滨海"爱勤道"，由"情节"至"情节"之间的道路是由绿树及灌木遮挡——使其不能一眼穷尽，始终保持游客对于"下一个情节"的期待感。

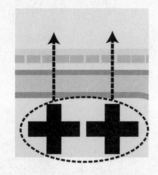

In the Sai Wan Ho community park, the path by the beach links all the landscape plots and recreational facilities systematically. One could get the experience of the sea from all the plots. Only by lingering along the waterfront, you will get a calm mood. The path from plot to plot is hidden behind the bush: you will never gain its full image by one look. This always maintains your expectation for the "next episode".

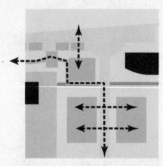

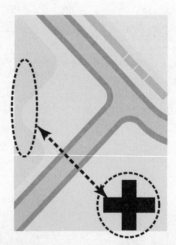

滨海公园以线性条带连接，每个点的元素将发挥不同的作用——使公园对于"住区"的贴面最大化，任意一栋住宅可以到达滨海公园的某个出口。
各种元素的偶然接近，都由不同的绿化分隔导致了随机性的"聚集"，赋予每个小块场地不同的"特性"。它们对于整个基地的"反射"作用创造了一种通过"片断"达成统一体的可能。这些"情节"分布的坐标是基于居民及造房者数量，以及使用频率而设定。

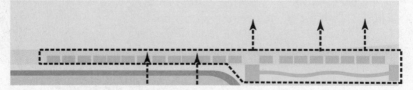

Since the beach park is arranged in a linear way, every element will play different role. The maximum interface toward the community guarantee the easy accessibility from any exit of the residence. The accidental meet with different elements results in a random connection by the division of the landscape plots. Each plot therefore is given its own feature.
Their reflective effect generates a unity through the "fragments", the locations correspond to the population distribution and the frequency of the actual demands.

ambiguity in public realm 04
公共空间特异化：鸟亭　（荃湾）

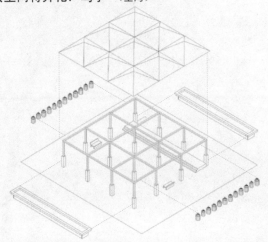

有一类公共设施，当其初建成时，其属性及功能的定义是中性的——无差别的面向所有大众开放，但是在实际中，却可能由于某一类人群的高频率或特色性使用使其变得相对专有化。
荃湾一座公共住区的"社区亭"，有16mX16m的尺度，由4m的钢框架柱网构成，上覆四棱锥玻璃顶盖，内部设有座椅，设计为开放型户外空间，供所有居民休憩与交流。无人料想，可能由于其钢框架结构是悬挂鸟笼的天然选择，亦或由于社区中老人是最长时的社区设施使用者，基于相同的兴趣，越来越多的老人将鸟笼悬挂于该处，这个"无指向性"的亭子成为了"鸟亭"。出乎意料地，它的使用专属度，排除了预设的中立性——但是它仍然向所有公众开放。
空间指向在平均化与特指化之间的转换，是香港公共空间非确定性的又一表现。

There is one type of public facilities, which is neutral when it's built, but might become specific through the usage of certain group of people. The community kiosk in one of the Tsuen Wan social housing neighborhood is openly designed for the residence to rest and chat. However, its structure is an ideal frame to hang the bird cages. And the aged people are the group who have retired and normally will stay at home. They are the main users of the kiosks. Based on the same hobby, there are more and more old men putting their birds here. Gradually, the kiosk becomes a "bird realm". The neutral is changed to specific. However, it is still available to all the residents and visitors.

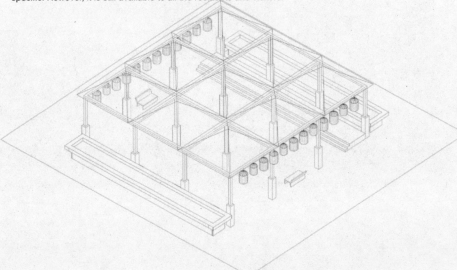

ambiguity in public realm 05
有时公共　（中环+湾仔）

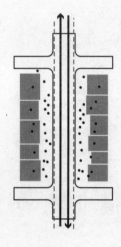 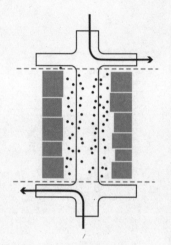

In Hong Kong, on the weekends certain vehicle roads in Central will be pedestrian only, this law is generated mainly for the assemblage of Philippines.
There are large amount of Philippine people who mainly work for family service in Hong Kong. Far from their motherland, they have strong demand to meet with each other on holiday.
However, in Hong Kong it is hard to find enough void for large scale assemblage. The government decide to leave some of the roads enclosed for them.

基于时间的公共性
公共性可以并不总是稳定的（或确定的），它可以随时间变化而产生改变。在开放空间极为有限的状态下，作为快速车道的道路在周末交通流量并不过分紧张时，被征用为市民的步行道。在一片只有很少"土地"可以使用的土地上，程式被交替地不断刷新，重新定义"公共领域"的含义。这种转化是动态的、临时的，以及"永远更新"的。通过限制某一类使用者，它鼓励了其他不同类型的使用。这至少说明"空间上的限制"可以通过增加另一维度——"时间"的方式来解决。通过系统化的管理，将纯粹的空间问题变得"开放性多义"，并为其设想了两条解决线索："拆散"和"组织"。

The public space is not always stable, it could also be changed from time to time.
The fast link only for cars could also be completely cleaned up for citizens' only. In a city which with quite limited "land" available, the "programs" continuously re-define the public realm.
The transformation could be dramatic, fundamental, and "forever new". By limiting one type of users, it encourages the other different type of using. Could the limits in space be solved by adding another dimension----time?
The purely spatial problem is transformed into the issue of "organization".
The beauty is in the "make-shift" strategy.

Ambiguity in public realm 06
延长公共时间：商业区座椅 （九龙）

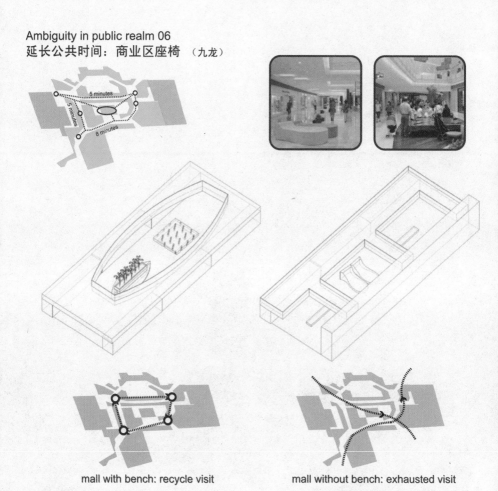

mall with bench: recycle visit　　　　　mall without bench: exhausted visit

细节中的人性
九龙站地下的购物中心"Elments"是香港少有的在购物廊道的所有沿途都设有各种舒适的座椅的商场。对照其他商业中心，基本上很难觅得可以免费坐下憩息的空间。商家不设座椅的理由是：免费的休憩空间会减少人们去商店内的餐厅及茶座的几。商家的逻辑是：如果你要休息，你就要消费——赤裸裸的绝对利益至上主义。可是，非人性的商业空间真的能提升餐饮类的消费么？还是仅仅是一种短视，更多的损害了消费者的热情和购买欲？Elments率先做出了尝试。虽然空间巨大，按照五种元素分成五区，流线也很长，但是由于沿途每不到三分钟的步行距离就设置了一组休憩空间，所以消费者在空间中不会感到疲累，在商场中停留的时间大大增长，商品买卖的成交几率也因此上升——它带来的潜在连锁价值是无限的。而由于人气的旺盛，商场内的多家餐饮店非但没有萧条，反而显出异常的旺盛。
对于所谓利益的过度短视，而不注重长远利益，会阻滞真正的商机。对于消费者来说，人性化的服务诚意才是更重要的。在现代消费社会中，购物已经不仅仅是购物本身，它几乎成为大众最重要的公共活动之一；买卖本身已经不是最重要的，它有更多的附加体验价值。

The definition of "archipelago" is a metaphor. It has two layers of meanings:
In the urban scale, it located in a deserted area, the only surrounding is nothingness, its only connection with the city is the infrastructure. In its unique confrontation, the strategy to stay far from the city centre rather than to be close, to isolate itself rather than "melting into" seems to be more rational here than the other way around.
The second layer refers to the fact that the luxury residence and the business skyscrapers forms "enclaves" of their own--only open to their own residents or business men, the public factor only exists in the shopping mall and the metro----even the sea view and the seaside space is also authorized. These generate the archipelagoes within the archipelago---for ever buffering for the flow----the inherent tortures of this kind of development.

ambiguity in public realm 07
公共集中化 （沙田）

public space and facilities

residential programs

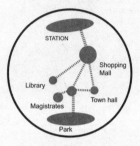

集中化策略
为了最大化使用和强化中心作用，沙田的主要公共设施全部在沙田地铁站200m半径范围之内。沙田作为香港为解决中心密度过高的问题而设置的新市镇之一，于1970年代开始建设。将市民中心和交通的中心并置于一处，赋予了沙田中心一种战略地位，成为该新镇社区的重心，这是一种资源最大化集中的策略，避免了新镇建成后容易出现的"低凝聚力"现象。对比与之同为新市镇的屯门及天水围，由于公共设施较为分散，交通与公共设施的结合欠缺连续性，预设的"自足城市"的目标并没有实现——人口除居住外仍然大量涌向城市中心，平时则成为一种"抑郁的存在"。

In order to maximize the effect of the "centre", all the public facilities in Sha Tin are located within the 200m periphery area from the city centre. Sha Tin is one of the New Town Plan triggered by the government from 1970s'. Concentrating all the public facilities around the transportation centre, gives the Sha Tin city centre with strategic importance. It focuses the resources maximumly at one point, with easy accessibility to all the surroundings, successively avoid the low attractiveness phenomenon which normally happens in the new towns. Comparing with Sha Tin, the same type of new town such as Tuen Mun, its "urban autonomy" in plan is not realized because of the weak connection of the transportation system to the public facilities. The population flow back to Hong Kong centre, the community only serves for sleeping.

ambiguity in public realm 08
公共空间潜力研究

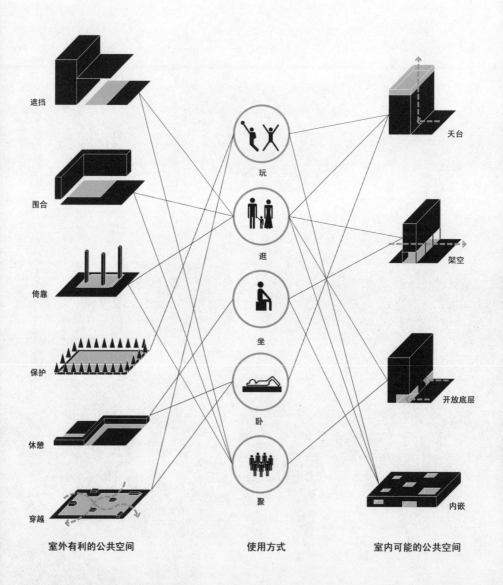

场地心理:可能的公共空间的场地必须是这样一种悖论的混合物:同时满足开放和私密的要求。开放是一种民主,必须使所有人有均等的机会进入和使用;而即使在公共场合,人在心理上也不希望自己的活动一览无余地暴露于众人的目光下。对于安全感的寻求使人们更倾向于使用可以倚靠、略带遮挡的空间。
场地选择:地价超高的都会中,除却公园之外,公共空间大多数指向"相对剩余空间"——活动只可能存在于某些特定区域,例如,架空的底层、天台,或者商家提供的部分室内。根据"自我感应"策略,民众的使用往往才成为场地所要激发的公共性的答案。

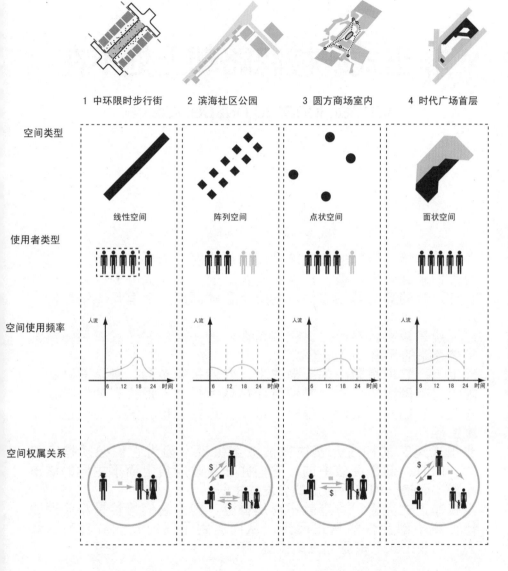

Locus psychology: The space which has the potential to be public must be a paradoxical mixture: to be both open and private. To be open is to provide everyone equal chance to access; to be private means that psychologically people still need some privacy even in the public. So they prefer the space which has certain level covering or providing a chance to rely on.
Locus choice: In the high land value metropolis, the public space normally relates to the "left over" spaces, such as the opened ground floor, terrace, or some interior space provided by the developers. The actual using will be the answer to the publicity which they want to promote.

5

非正式：被忽略却无处不在

Informal: ignored but pervasive

香港通常给人这样的都市映像图景：华丽、繁盛、巨型、高耸。当人们来到香港之前，想象中的香港是一个摩天办公楼和银行的丛林，有闪亮的幕墙立面，层层堆叠的高架路上有川流不息的车流……

这无疑是事实。然而，这仅仅是香港的一面；它还具有另一类常常被忽略的元素："非正式"。

尺度上，它们是微小的；形式上，它们是简单的，甚至丑陋的，因为没有确定的形式，它们可以被认为是"无形式"的；存在方式上，它们极其灵活、随机，可以在正式城市无法存在的任何土壤生存。

非正式建筑的构成元素往往被减至极少，因为它们必须是廉价的，同时，却是高效率的。它们并非是政府自上而下主导的城市元素，而是一种自下而上的自发式生长。

没有非正式元素，香港就不再是香港，因为它们恰恰是将亚洲式特征赋予都会性的真正因素，从而区别于任何欧洲或者美洲都会。"非正式"是亚洲性的灵魂。

The city appearance in Hong Kong always gives the people impression of majestic, magnificent, massive, lofty and beautiful. Before people come to Hong Kong, what exists in their mind is a city with skyscrapers offices and banks; shiny facades with glittering elements; crossing overpass with never-stop vehicle flows........

However, there is also another kind of ignored elements: the informal. In terms of scale, they are small, tiny rather than huge and monolithic; In terms of form, they are random, simple ,even ugly, because of no fixed form, they could also be considered as "formless"; In terms of existence, they are super flexible, can grow in any soil where the formal city elements become incapable.

They appear everywhere you want them to be. The facility and instruments of informal is normally declined to the minimum level to be cheap, at the mean time, they get the maximum efficiency. Instead of the top-down city from the government and the urban planners, they lead their own bottom-up logic of growing.

Hong Kong can not exist without the informal, because the informal is the real entity which identify it as Asian metropolis, which differs it from American or European metropolis. They are the ghosts of "Asianness".

informal 01
阳台重构：个体自足（荃湾）

为获取更多的个人可支配空间，高层公寓的居民用各种方式"重构"阳台——基本方式是"封闭"，使其成为"室内"空间的一部分。但是，由于各个独立住户使用的目的存在差异，他们封闭的手段、方式以及选材都彼此互异。窗户的尺度、分割、全封闭或半开放、护栏的高低、有无空调窗机……都成为使差异加剧的因素。

从政府的立场来看，未经许可的"自建"其实是一种不合规范的消极行为——它制造了立面的混乱，影响了市容的完整性和整洁性。然而以居民本身的立场看，这种"自发围合"完全来自于过分逼仄的生活空间的压力所引发的个体对于现实的一种无奈反抗——空间的不充足感已经紧迫至生理层面：切肤的、憋闷的、难耐的。从建筑的角度论断，至少公屋中普遍存在的现代主义的标准化立面被"自发建造"打破了，"平均化"成为"个性化"，"对比、多样"取代了"简单、统一"，这是一种集体无意识的"最大化混合"运动。似乎令人难以置信，仅仅是阳台的利用就孕育了一种独立于建筑师意志之外的无限可能的条目。

In order to gain more individual space, the residence use all kinds of methods to reconstruct the balcony space, the basic logic is to enclose it, make it part of the interior livable space. However, because of the difference in aim of usage, the way they enclose it differ from each other in detail.

From the government's view, it is negative and illegal: it breaks the neatness and cleanness of the city appearance. But from the residence own positions, this kind of self-built phenomenon is the representation of their actual wills to gain space.

At least, the original facade autonomy is broken by the non-unified elements. The standard becomes uniqueness, the contrast replaces the unification.

It's a "mix-max" strategy by the auto-collective movement.

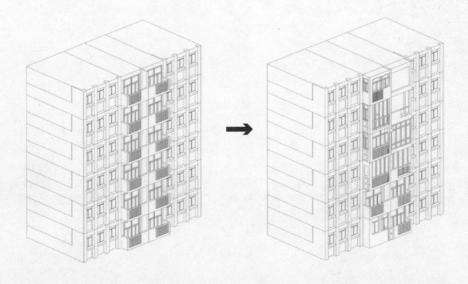

informal 02
立面重分　（荃湾）

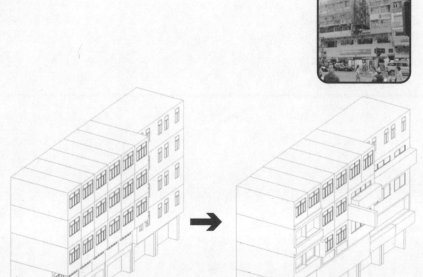

由均质向特异的转变
建筑，在观念内通常被认为是一种不易改变的实体，无论形式上或功能上——从古至今，它总是与确定、稳固，牢不可破的客体性概念等同。但是，在现代商业社会中，尤其是香港这样"无所不可能"的城市里，这种定理似乎不再绝对。基于利益的驱使，建筑可以在形式上和程式上都发生戏剧性变化。荃湾建造于七、八十年代的公共住宅，由于年代久远，建筑生命的自然衰减已经令其不再适合居住。但是都市的扩张和发展往往没有人可以准确预知，昔日的住区由于临近地铁站如今成为火热的商业街，"过时"的住宅因占据极佳地理位置成为商家的争夺热点。不同的商家根据其商业所需营业尺度向房主高价征用不等面积的房屋，并且占据了不同的位置和单元数量。标准化的现代主义住宅经历了"重划分"，成为各种奇异空间的组合，而立面也因为商家的不同色广告成为"城市拼贴"。
这种现象有趣之处在于：均质立面成为"程式化的立面"，你可以从立面的划分直接"阅读"出其中的程式及空间分隔。

From the generic to the specific
Architecture is one solid object which is supposed to be non-changeable either formally or functionally. However, in the modern commercial society, this theorem seems to be not absolutely true. Driven by the profit demand, the building could alter dramatically both in program and form.
In Tsuen Wan, housing built in the late 1980s' and early 1990s' is now in bad condition because of decay, not suitable for living. However, with the urban expansion, the formal housing area now becomes busy commercial streets. The "old fashioned" residence therefore is highly valued for street shops.
The house owner rent their houses to different stores and brands. Different store developers occupiy certain part of the housing, according to their business scale. Gradually, the standard modernism generic housing units is re-divided into all kinds of realms, the facade also becomes a " collage".
The interesting thing here is the facade is a "programmatic facade", you could "read" the program and space division through the facade

informal 03
桥下填充 （铜锣湾）

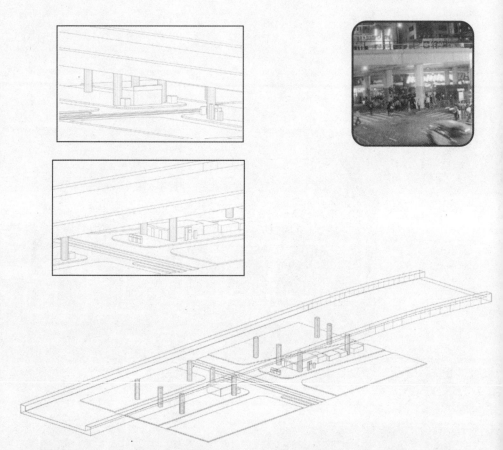

自动占据
如果我们采用某种最小化控制、最大化鼓励民间智慧与源动力的方式进行城市更新，而暂且把过多的关于"紊乱"的担忧搁置，将会产生怎样有趣的城市元素与现象？

高架路下的空间，遵循"城市填空"原则——一种几乎独立的潜在自由，以城市新增维度的方式进行表达——可将其转化为一种实用主义的发明么？如果我们认同海德格尔对于建筑原形的定义——"棚子"，那么"桥下"无疑是这个定义的最好表达：有顶盖遮风挡雨，有立柱作为支撑，容纳各种活动。虽然因混凝土的粗野与交通的噪音使它不适合做正式建筑用途，可是对于非正式来说却是极佳的存活土壤。铜锣湾坚拿道西的与轩尼诗道的交汇口处为高架，其下不仅有街坊福利会，还有一个条状街道公园。这是本世纪之初仍然保留的自发方式对于上世纪后半叶非谦逊式城市建设的回应与补遗。

"Auto conquerism"
According to Heidegger, the "Prototype" of architecture is simply a "shed": four wood columns holding a roof by grass. If we admit that it is the entity of the "thing", the underneath of lifted highway is an ideal architectural space. Its emptiness provides a lot of possibility for accommodation of programs, with a natural roof to protect the under-world from rain and sun. According to the principle of auto-conquerism, this kind of space is suitable, (even already luxury) for new insertions.

You could find all kinds of citizen facility here, such as street welfare house, street park with benches, kiosks etc, covering all kinds of daily life.

informal 04
"粘贴式"宅前店 （湾仔）

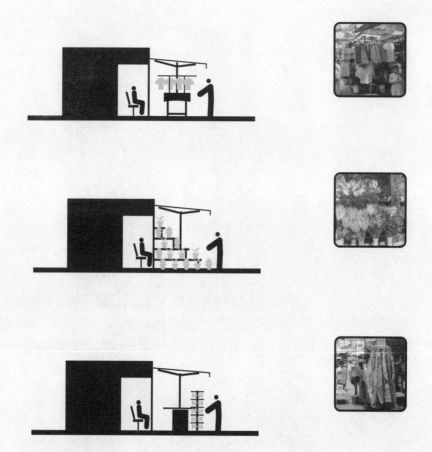

非正式宣言："我们的角色并不是在当代都市中被摩天楼挤得无处藏身，而是使冷漠的摩天楼都市变得更加人性。"

湾仔的个体商业经营者发现了一种远离生活悖论的方式：一种将商业意图与居住时而结合、时而分离的技术，它依赖门前可撑开与收起的屋架的灵活性、非正式货摊的简单性及对于"模糊性"的操作。当使用这种"附加媒介"进行商业活动时，一种将功能拓展的技术，一种在白天将自家前厅变做半公共商业面积，在夜间重新将其转化为私密居住空间的解放得以实现。单一空间在双重功能的矛盾压力下被反复重构。

The house which is located at the commercial street front is transformed into commercial use as well. By removing the front door and installing a simple canopy on the top, the residence is changed to a mixed use: store in front and living at the back. What's interesting here is the semi-public space created by the individuals, the absolute division between public and private is blurred.

informal 05
可折叠商铺 （旺角）

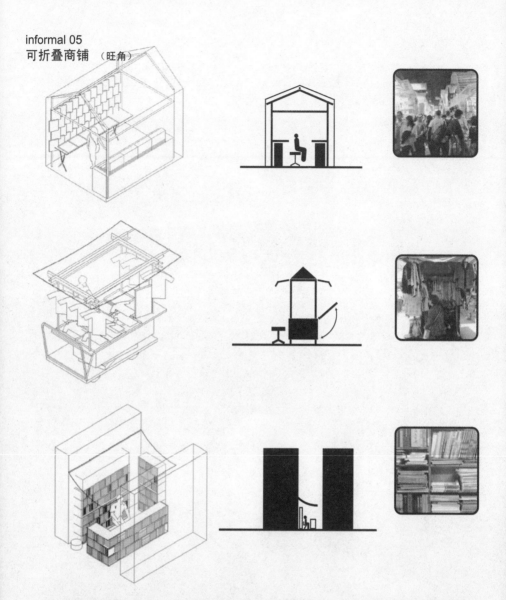

非正式是对于"最小化建筑"的民间自发式探寻——形式上以及功能上，正在将城市的刻板界面以一种充满人情味的市井方式取代，将封闭、刚性界面转化为柔性界面。这种探寻的深层次落实于心理层面。

几乎是同时又是反向地，香港在"正式建筑"之外选择性地保留了非正式作为其城市的另一种养分：城市自身已经感受到纯粹的自足性摩天楼引起的营养不良带来的阵痛。非正式是都市状态下注入的一种村镇式生活，如同液体一样流淌在都市巨石的每个罅隙之间，并且生机勃勃地与之共存。也许永远都不是绝对"纯都会"的都市，才是都会延续的最终宿命。

Could we imagine one building which could be generated from "nothingness" to "something"? From flat to Volume, and also could disappear also just within 10 minutes? Large amount of the "informal building" could be foldable. Because of the informality, they have to search for all the possibilities to survive. It can be in the gap between two high-rises, (the distance is even less than 2m), it could be a minimum convenience shop, a book store or a lock shop. They will be simply built up with simply some cloth, iron sticks, and tables in the morning, and decomposed again at night just in 10 minutes. They are everywhere, They are no where.

informal 06
伞筑 （中环+湾仔）

伞的异化
一把遮阳伞与桌子的结合，可以成为售货亭；一把伞与长凳的结合可以成为遮阳座椅；几把伞与货框、货柜的结合可以成为小型市场。它的妙处在于无尽的可能性——对于两种事物的功能的简单嫁接，产生出完全新的事物；在于合并之后的整体的效用，是合并前的任何一个部分个体都不可能完成的，属于1+1〉2的模式。这种操作是物与物的嫁接，更是两种概念的杂交，单纯的拼接可以产生非单纯的结果。它预留了一个空白，不断提供新的可能。

How many things could an umbrella do?
One of the main features of "informal" is the flexibility. This can be shown through the use of umbrellas. Combined with a bench, the umbrella could be a kiosk; Combined with shelves and tables, the umbrella could be a store; Several umbrellas tied together with fruit boxes, you could get a "Mini-fruit market".
In different occasions, the umbrella plays different roles. The creation could be achieved by "switching concept", it redefines the meaning of building, and gives inspiration for us to make "newness".

informal 07
误用与转用 （深水埗）

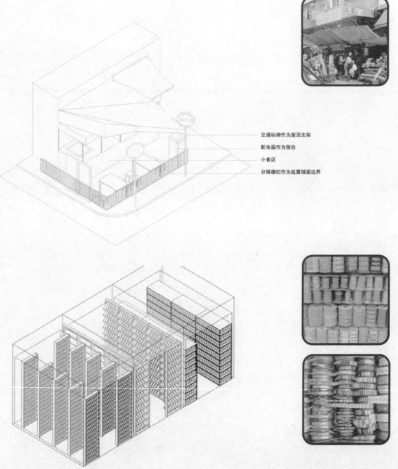

交通标牌作为屋顶支架
配电箱作为围合
小食店
分隔栅栏作为延展铺面边界

误用、占用与转用
"非正式"还常常表现为私人对于公共或半公共资源的占用或误用。基于自发的使用需求或者利益需求，将公共空间的一部分纳入自己的版图，或者将本具有某种公共用途的物品转为他用。例如深水埗的一家位于街角的小食店，在步行道护栏范围内摆出桌椅，将其纳入营业的一部分，并且利用标示牌作为屋顶的支架。这一类"占用"得以存在的条件是：该物品或领地本身具有天然的可利用性，具有多种用途的潜能，并且被夺取的某种公共利益必须以另一种方式返还予公众。因此小食店虽然占用公共资源，但是仍然受到民众欢迎。
而转用则表现为对于某种熟知物体的"异常"使用，是一种对于"功能"的再开发。与误用不同的是，这种方式仅限于私人个体自身空间的灵活转化，并且，也只有在某种特定的条件下才会产生。仍然以深水埗为例，这里的汝州街是著名的布匹及饰品批发市场，由于品种繁多，为了便于展示，所有的门面都被当作玻璃橱窗，由各色样品将其无限细分，呈现出一种"目录化的倾向"。这时，立面已经不仅是立面，它等同于一张"货物清单"。

The informal is also shown as the misuse of the public resources. Based on certain autonomous demand, the individuals include part of the public space into their own, or make use of some public owned furniture. The snack shop owner in San Shui Po put the tables onto the pedestrian road, and use the transportation sign board as their roof stick,
The other type of informal is trans-use of certain objects. This way is limited in the private property itself, and only happens in certain conditions. For instance, the cloth shops hang their samples of good on their façade, the façade changes to a showcase, and it is categorized façade.

informal 08
非正式类型研究

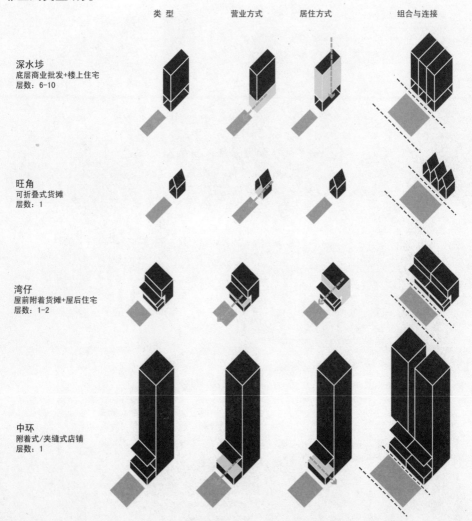

非正式的存在方式差异,体现在对于"场地适应原则"最直接和字面的贯彻。传统货摊式、市场式的经营方式对于环境差别的无挑剔性,已经放任了其在都市每个角落蔓延的滋长能量——甚至是都市最现代和规整的区域——以自身的可变适应周边的变化。如果说作为"图像意象"的地域性(例如大屋顶建筑)已经在亚洲不可逆转地逐渐消失,令人倍感无奈又无可作为,那么,非正式却以"程式的亚洲性"顽强地生存在都市的坚壁之中。这种不属于形式或符号范畴的地域性,却切实存在于东方生活的各个层面(甚至是意识层面)。奇怪之处在于,无论东西方学界,对于"符号地域性"关注备至,而对于"程式的地域性"却漠不关心、视而不见。

非正式以其略带狡黠和粗鄙的方式,根植于对于惯有生活方式的热爱,与"正式建筑"一同实现了对于"亚洲现代性"的宣言,无关任何建筑潮流或手法。

The existential difference of "informal" is revealed by the very direct and literal conduction of the "adaption to the locus principle". The traditional market doesn't rely on certain site. This has liberate its energy to expand to any corner of the city——even the modernist and neatest part. They suit to the surroundings with their flexibility. If the visual regional identity of Asia has been removed by the modernization, the programmatic Asianness still grows in the rock of the modernity. This identity does not relate to any image or symbol, but exist in the urban life of Asian cities. However, this kind of actual identity is always ignored by the critics.
The informal, rooting in the passion of life, has realized the manifesto of "Asian modernity" in a crafty and rough way together with the "formal", but without any relationship with architectural tendency.

6

效率最大化

maximum efficiency

高密度城市的运转要点是：高效的基础设施支持。
当人口超量聚集，密度急速升高时，在同一时间内，各种流路（人流、车流、物流）都必须以最快速度到达、通过、及疏散。物质的聚集必然导致各种流路之间的交叉冲突；通道容量的限制，激化了这一冲突。高效的基础设施必须承担清理、分流及快速传导等多重功能，使各种机能协调运作而不互相干扰。
在这里，你将看到因密度而催生的交通层次，向空中和地下正反两极同时延展；你将看到天桥和廊道所建立的平台，使两种不同速度和使用需求得以完全分离，成为两个世界。
在漫长的城市自我调节旅程中，各种运行方式被实验；每个个体如离子一般游移，但均能快速定位自己的位置，赶上通往下一个目的地的输送器。

The key factor for the high density city to run smoothly is the completeness of highly efficient infrastructure. While huge amount of population condense and the density get sharp increase at the same time,
All kinds of flows (pedestrian, vehicles, objects, information) will have to arrive, transport, and be evacuated as fast as possible. The aggregation of flows will also lead to the conflict between different routes. The limits in the capacity of the transportation joints exacerbate the conflicts.
The highly efficient infrastructure must function multiply, to sort out, to divide and to transmit the flows; to make the city organs work in harmony without interference to each other.
In Hong Kong, You will see the transportation layers inspired by the density, expanding simultaneously upward and downward; you will see the terrace established by the overpasses separates the city users into two worlds according to the speed.
In the long journey of city auto-adjustment, all kinds of operations have been experimented. All the people and objects move as anions, however, they could easily find their own position, and catch up with their next vehicle to the next stop.

maximum efficiency 01
负密度：香港地铁（一）

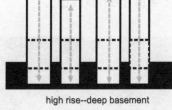

EU CITY   HONG KONG

thin basement, short distance   high rise--deep basement

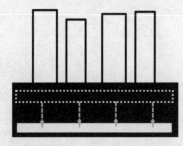

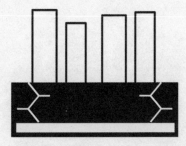

the metro could only be deeper underground   long distance, long circulation

亚洲都会的高密度发展，不仅仅包含指向天空的成簇的摩天楼，它同时也是向地下的延伸：高密度永远是向正负两个方向同时发展——几乎是一种牢不可破的镜像。向上的高密度必定有向下的地下基建设施与其呼应。随着建筑层高的加高，地基深度也相应更深，那么地铁只有往更纵深发展。我们可以称之为：负密度。反观欧洲，其低密度的城市形态也导致了地铁的浅层化。这验证了在城市密度增长量方面，正与负的发展是成正比的。

The high density development in Asian metropolis, is not only the skyscrapers toward the sky, but also something toward the ground; It's not only the city exposed in the sunshine, but also something deep hidden in the darkness; It's not only the above zero part, but also something minus.
And they are like the twins who rely on each other, the higher density you get above the ground, the same depth you have to achieve under the ground.
It's a mirrored development process.

maximum efficiency 02
负密度：香港地铁（二）

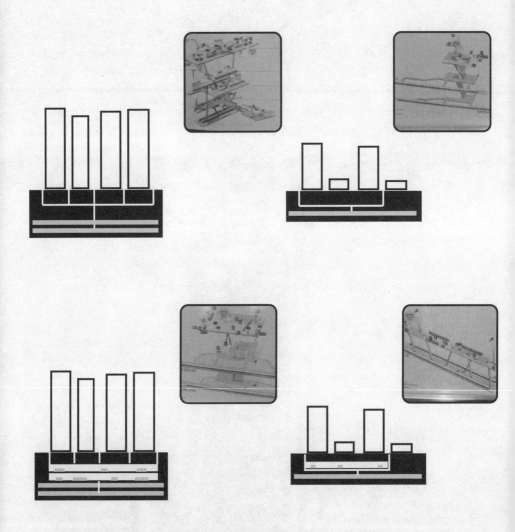

地铁作为一种现代都市交通的载体，已经极大程度上确定了建筑之外的城市基础设施主导城市生活的现实。作为有确定地点、相同的频率、无论在时间还是空间上都极为精确并有规律的系统，地铁稍纵即逝的特性维持着亚洲都会的城市血液的输送。
地面上密度越高=地铁向下挖掘越深=地铁站将提供相应更复杂和完备的服务设施=更多的出入口，路径以及与地面更顺畅的联系=更长的流线——制造更多与商业界面贴合的机会。

The higher density you have above the ground
= the deeper the metro station dig into the ground;
= the more complexity and completed service facilities the station has;
= the more exits, paths, and wider connection with the ground level;
= the longer circulation you have, the more frequency to meet with the commerce.

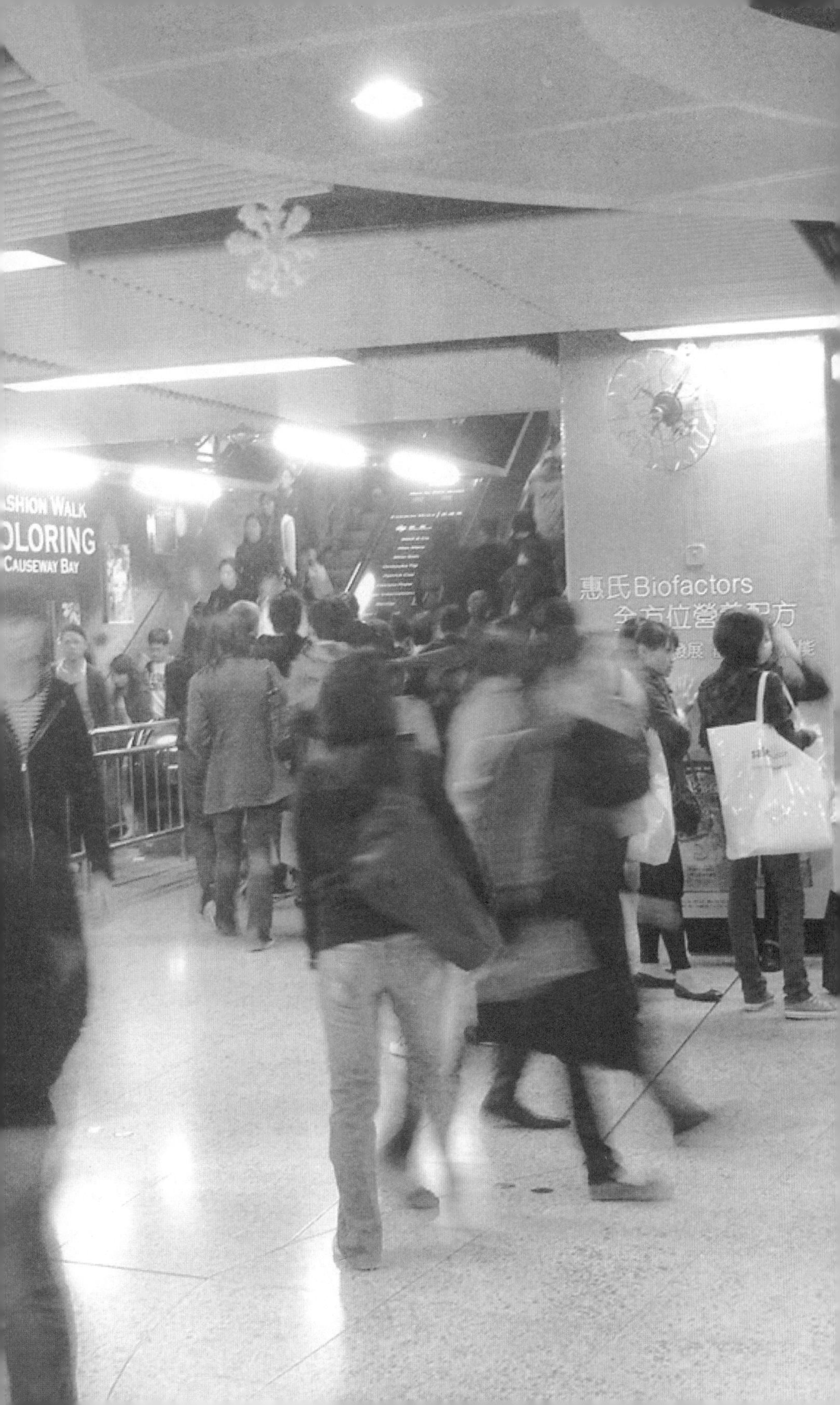
CAUSEWAY BAY
惠氏Biofactors

maximum efficiency 03
负密度：香港地铁（三）

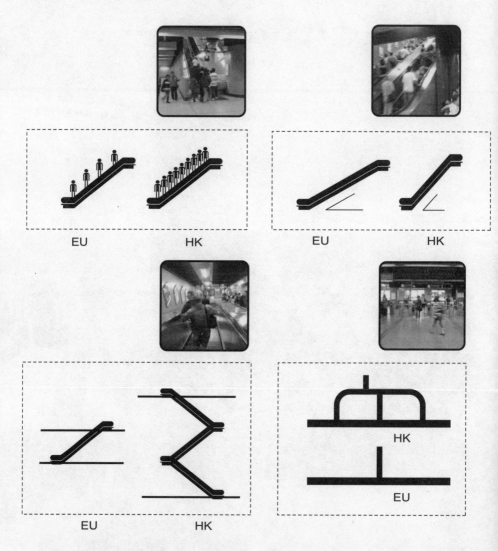

即使是已经确定的城市运转的难度——在最短时间内将大量人流快速搬运——在以欧洲低密度城市为参照物的体系中，这种效率成为香港地铁乐观主义的基础。根源性地，是在资本驱使与市政努力的双重鼓励下实现的。在都市难以想象的黑暗地底，都市生活正在不止息地以人工方式循环。它挑战了一种存在已久的偏见：高密度一定等同于低效率。香港地铁以5倍于欧洲地铁的频率运行，单位时间内运载了10倍的人流，它的效率是欧洲地铁的50倍。

Comparing to the metro in European cities, the Hong Kong metro is several times deeper, and it has to transform 10 times more people in more frequency. The amount, capacity and speed of the escalators also have to be much higher.

The interesting thing is, with higher transportation density, with more people to be transformed in shorter time, the supposed chaos never happens. Everything goes fluently and orderly, the people move faster and smoothly. It challenges a long existed dogma: The high density definitely relates to low efficiency.

maximum efficiency 04
负密度：香港地铁模式

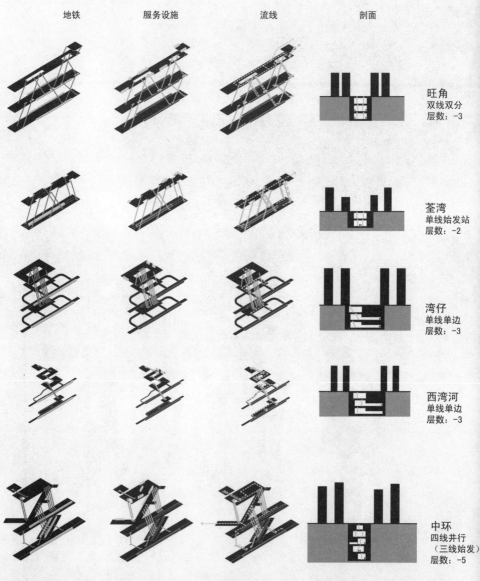

香港地铁是对地面状况和场地价值的最精确对应。地铁的管线、轨道、方向、层数，几乎形成一种对于"需求"的图表式诠释，一种完全为"流动"而生的极简主义，毫无任何冗余。通过地铁，你甚至不出地面，就可以直接预知地面上是"中心"还是"边缘"：地段越佳，理论上地铁站就有复杂化的趋势。在某种意义上，通过对于均质的高强度流动性与人群行动的非确定性混合，整个地铁是一个基于时间的空间系统。这种建筑摒弃了所有美学要素，是功能性被放大至极致的产物，它与机场有相同的特质。

Hong Kong metro is the very accurate reflection to the situation and the property on the ground. The tunnels, the track, the orientation and the layers of the metro station almost form a diagrammatic interpretation of the demand: a minimalism for flows. You could tell the happenings above the ground even if you are still underground, where is the central and where is the edge of the city. The metro is a spatial system based on time, it abandons all the aesthetics, but enlarge the functional aspect to its extreme. It is similar to the airport.

maximum efficiency 05
"无水的威尼斯"：天桥系统（分流）

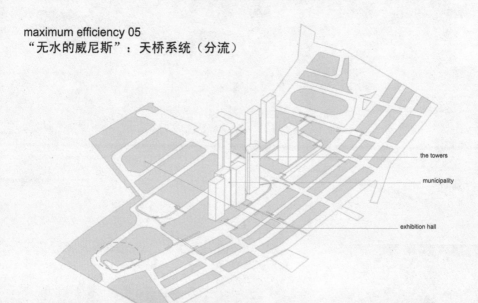

the towers
municipality
exhibition hall

高架人行道在一种都市尺度上，以其直接性、顺畅性以及连续性提供了"穿越可能"。两栋建筑之间以点对点的方式连接，一列建筑之间以线性连接，而多栋建筑之间则形成网络式面状连接。它是一种开放式的更新，对于深受交通阻滞之苦的居民，它提议了一种将步行与车行分离的可能：远距离城市旅行以车为载体，而短途或局部城市旅行则可享受步行的悠然乐趣。

从湾仔会展中心，经过商城，经过民政大楼，再至地铁站，所有市民常用公共设施被提升至二层的，由有顶盖遮挡的天桥连接，市民可以毫无阻隔、无须停留的穿越。兼具最大可达性及便利性，同时为地面提高效率——一种双赢的策略。

The bridge linked all the buildings from the metro station until the Exhibition Hall, though municipality and shopping malls without any stops. It's a non-obstacle connection, with maximum accessibility and convenience for the pedestrian, also simultaneously regain efficiency for the vehicle road.
In this trip, you will experience series of episodes, such as culture buildings, commercial and landscapes.

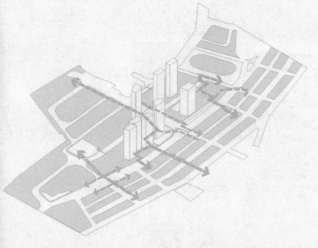

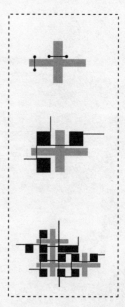

maximum efficiency 06
"无水的威尼斯": 天桥系统（网络）

城市地块的价值因其"连续性"而进一步提升，各个部分更多纳入一个整体同时运作——相应地逻辑应运而生：想要获取更多机会，就需要向这个既存网络靠拢；新的成员不断被吸纳，天桥的延展度也成为无限。
——一种新的，由廊道定义的建筑生长可能（生命性）；
——一种在私人办公楼丛林中保留强烈公众可参与性的氛围；
——一种依据法规可使其网络不断延展完备的自足裙房系统。
两种类似生物有机体的精神被赋予给无生命的城市人工物：1 网络不断扩大，几乎无可控制，也无须控制2 公共体系以一种自我生长的方式进行扩张，如同核反应的链式效应。
虽然廊道连接的都市是一个复合系统，然后在整个都市体系下，仍旧只是部分。廊道连接的都市片断代表了一种对于香港步行公共空间的总结，显示了所有的策略、理论样式及潜力——以一贯"香港式"的简单直接的方式解决。

The plot value is increased by the continuity of the landscape, every part is operated within a whole system. The more opportunity you want to gain, the closer you have to stay with the network. The new members are continuously absorbed in, the propagating of the overpasses never stops.
1 An architectural forever growing mode generated by overpasses;
2 The mood to maintain the publicity in the private business area;
3 A forever refreshable podium system within the local zoning regulations.
Two biological effects are planted into the body of urban artifacts: 1 the endlessly expanding network, incontrollable, also no need to control; 2 the public system grows autonomously as the nuclear reaction.

maximum efficiency 06
"无水的威尼斯":天桥系统(连廊)

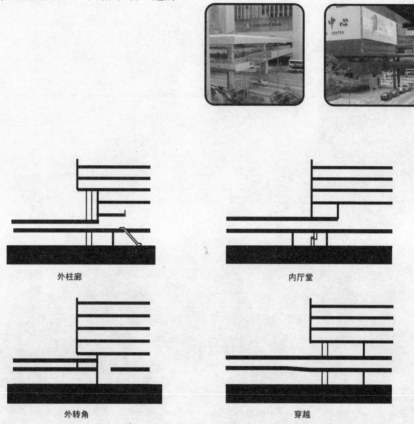

外柱廊　　　　　　　　　　　　　内厅堂

外转角　　　　　　　　　　　　　穿越

与其他任何城市的过街天桥不同,香港的天桥除了固有的"穿越"属性之外,尚有一个更重要的功能"连接"——由于室内购物广场及休闲设施的存在,天桥连接了地面与地上、地上与地上。当利用简单的"通路"来刺激香港消费主义的城市表皮时,其所有的城市孤岛都被激活——将各个独立的裙房自治体从孤岛状态下解放出来。这使任何业主在定义其开放地块功能时,都可以从"拓扑"角度思考其个体定位,避免过度竞争。

连廊连接裙房,使这个连续的室内空间不仅是一个连续的容器及输送通道,不仅是一个仅供"经过"的场所,而是可能催生连续变化的事件的装置。以其自我调节功能进行不间断的发展。连廊连接的场所成为一种舞台,每个人在其中都同时既是表演者又是观众,这里充满了"偶然"与"即兴"的发挥——空间与人的关系得到空前的凸显,行走不再是离散的过程,而是一种互动的空间文化。

The way of bridges connecting to the buildings verifies a lot from each other, depending on their aims. In other words, the connections show the gesture of how the owner of the building want the people to get into their land, and how to pass through, it generally gives such kind of information: welcoming, defense, guiding, or luring.

The one at the pacific centre have bridges to all 3 sides of the building, introduce the people into the main shopping mall; the bridge link the metro to the municipality in Wanchai directly leads the flow into the reception hall of the office, showing a kind of openness to the public. While the corridor bridge at the gate to HSBC headquarters only plays the role of "passing by", so although the bridge runs very closely to the facade, they almost has no connection to each other: the bank is not a public or commercial space.

maximum efficiency 07
"无水的威尼斯":天桥系统(可达性)

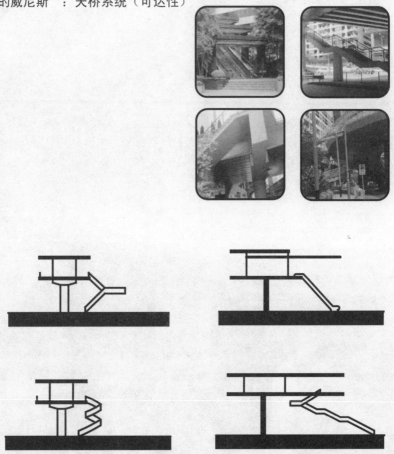

根据功能、人流量、连接对象及效率需求,过街天桥的通路、方向及抬升方式彼此互异。它直接回应了人流的行动目的,以及适时的"降落"、"升起"或者"连通"的欲求。楼梯的形式昭示了其功能。主体本身作为部分来展示,显示了主体行动的必要性。高架的通路、连接城市步行平台,它们一并形成了一种"现代的无水的威尼斯"。哈维.柯柏特在上世纪20年代年代为纽约提出的解决交通拥堵的构想在纽约未果,却在半个多世纪以后在香港大行其道:这是一个多个大洲意识形态多次受孕后的产物——灵感来源于欧洲古城,在美洲被提出,最终被亚洲的实用主义落实,赋予了终极存在。

According to the function, the amount of flow, and the efficiency, the sides of the path way and the type of the stairs could differ from each other. They correspond to the demand of the flow, and the requirements for "landing", "rising" or "connecting". The form of the stair exit/entrance shows the relationship with the function.
The subject itself is revealed as only one part, shows the necessity for the subject to act. They formed a kind of "Modern Venice", which had been proposed but never realized by the Manhattan theorists.
 Harvey Corbett: the roads are liberated as river, the cars are the "modern Gondolas", while the people-
------living on the bank passing through the street by all kinds of "bridges".
The road becomes a river without water, no water but the same mechanism.

maximum efficiency 08
天桥系统比较研究

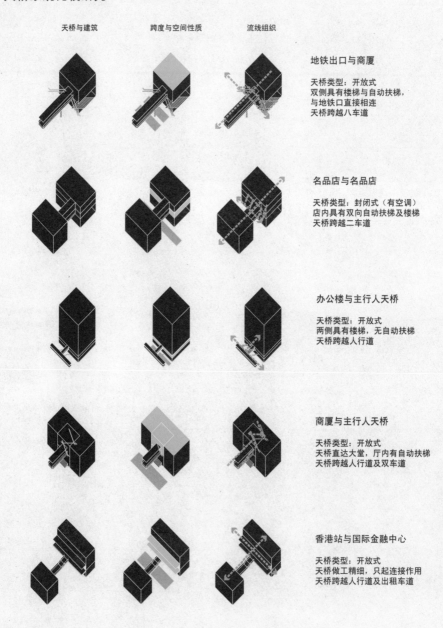

maximum efficiency 09
双层：可移动建筑

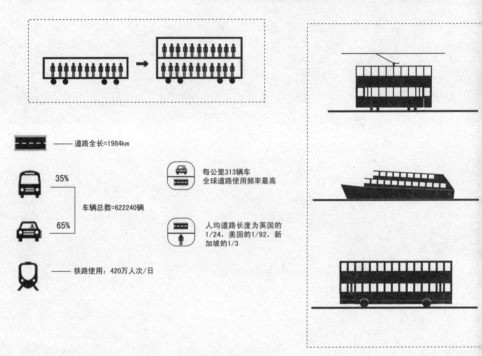

道路全长=1984km

35%
65%
车辆总数=622240辆

每公里313辆车
全球道路使用频率最高

人均道路长度为英国的1/24，美国的1/92，新加坡的1/3

铁路使用：420万人次/日

如果我们将交通工具视为某种可以移动的建筑，即可以理解为什么几乎所有香港的交通工具都是"双层"的——任何其他城市都可能有双层交通工具，但是如香港这样占绝对比例的，天下只此一城：有轨电车、巴士、甚至渡轮都为双层——如果技术允许，港人一定会把地铁也变为双层的！20世纪60年代前卫运动中的建筑电讯派（Archigram）对于"可移动、会走路"的建筑的构想并未实现，随着信息化的推进，人们发现建筑本身作为容器并无随处移动的必要，而对于交通系统的依赖却在高密度城市中空前加强。或许"双层"现象是"建筑电讯派"的另类实现：可移动的容器。

双层交通的贡献在于将密度由静止转入"流动"状态，使城市自足体可以大量吞吐其使用者，并将"高密度"变为一种大众习以为常的"日常状态"——将通常意义上的非常状态转化为"寻常"，使都会民众在心理上达到普遍的高密态。于高密度状态下，人群需要持续将事物推动至某种不寻常的程度——几乎是在无意识状态中，他们已经在进行对于"可能性"的实验，重新定义"极限"的含义。

If we consider the vehicles as a sort of movable building, we could understand why all the transportation is double layered: Double bus, Double tram, Double ferry, if it's technically possible, I guess there will also be Double Metro! The high density society need a more efficient transportation system.
In a high density condition, people need to continuously push the things to some unusual level. Unconsciously, they are always experimenting all the possibilities, to redefine the "extremes".

本节数据来源：香港政府统计年报

maximum efficiency 10
都市网格：最大化可变混合 （旺角）

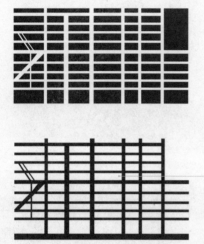

确定结构中的非稳定性

旺角至油麻地的都市网格呈现出与曼哈顿相似的结构特制：均质、平等，地块被间隔一致的网格以无差别的方式重复分割。这种城市现象在香港并不多见——香港一直以地块划分的多样性为行动原则。形式相似的背后也隐含了类似的逻辑——曼哈顿都市主义的精髓：在看似简单的都市框架下，结构的确定性与程式的非确定性完美结合；以最单纯的结构，包容了最复杂的都市可能。这个系统的突出优势是结构中单个地块上的内容可以轻易改变，而不会影响都市的整体布局，同时，它达到了程式与类型的灵活性的最大化。

The Urban structure of Mong Kok reveals a similar feature to Manhattan Grid, which also follows the spirit of Manhattanism: with the seemingly simple urban structure, it combines the plot definiteness with the programmatic instability. Within this system, everything on the plot could be modified easily without changing the whole layout; simultaneously, it also achieves the maximum program and typological diversity and flexibility.

7

边界状态

border condtion

亚洲都会的特异性还体现在：无处不在的边界。
边界以多种形式出现，它可以是具体的、可见的物质分隔物（如墙、樊篱），也可以是隐喻的界限（例如分道线、场域的交界），甚至可以是不可见、无形的（例如监视摄影机）。
边界具有分隔、围合、展示、隔离、控制、激发等多种效应。可控制空间，表达社会心理。所有的边界都是相邻界面双方阻隔意愿物化的结果，当双方内在的张力不足以消解这种对立时，实体边界便会干预其中。
边界可以存在于小至相邻的个体之间，大至交界的种群以及各种差异群体之间的相互对抗与自我保护，它区别等级、判定身份、划分族群，
在香港，处处都可发现现边界的设定（分界），也时常存在对于边界的突破（越界）。边界无优劣之分，却有消极和积极的影响。研究各类边界，可以找到破解城市密码的罗塞塔石；而通过削减消极或设定积极边界，同样可以实现对于城市的有效干预。

In sight into the entity of Hong Kong, there are all kinds of borders, physical or metaphorical, visible or invisible, They are generated by the difference in social class, manner, benefit conflict, This is one of the unique feature which identifies the Asianness for Hong Kong.
The borders divide, protect, enclose, provoke or represent the very inner psychological unspeakable factors of the city.
The border effect could happen in any scale, from neighboring individuals to large social groups.
Reading the borders, we could find the clue for the mechanism of modern society; by diminishing the negative, or setting up the positive border, we could also give intervention to the city more precisely and effectually.

border condition 01
新、旧与过渡 （西湾河）

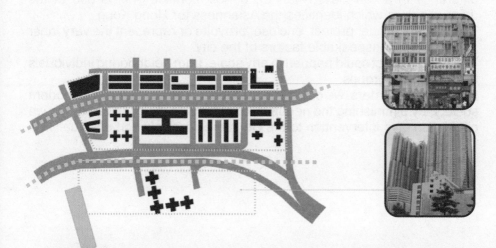

阶级的边界
旧城区与新建住区分属于不同的都市时段。以西湾河为例，升起的高架路及地面快速车道成为新城与旧城之间两道天然的分界线。建筑类型、街道网络及相关的生活方式在边界线两侧形成鲜明对比：南侧为拥堵、破旧的老区，而北侧为临海、整洁的开阔住区。两道基建之间的地带为过渡地带，呈现出新老混合的多样状态。实际上，隐藏于现象背后的、由社会等级的差异所造成的"边界"比实际的物理分割的效应要强烈许多——由心理上的领域感造成——这种边界不可见，却更加有力。

The old urban district and the newly developed estates belong to different urban time. The lifted highway and the express way become the natural border of the new and the old. The building style, street typology, and the related life style form sharp contrast on the two sides. However, the social class difference hidden behind is more profound than the physical urban appearance.

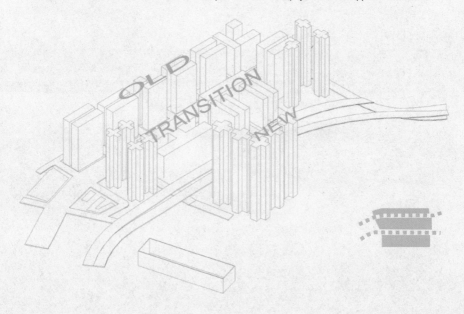

border condition 02
场域边界 （铜锣湾）

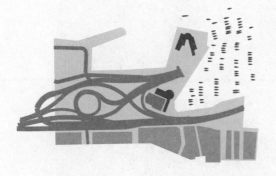

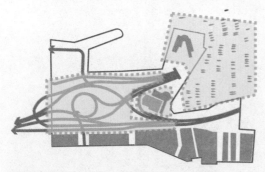

香港都市中有一类边界，表面上不可见，却实际存在，并且正在切实发生作用。不可见是因为并没有物质概念上的阻隔，如墙或樊篱，它的边界效应来自于"场域力量"。
铜锣湾的近海海湾区设计定位为"公共区域"，实际上却存在许多"匿名"的圈地。从铜锣湾商业区向海滨区进发，首先要穿越有高速车流和大量噪声的快车道及高架路，加上人行天桥高低起伏和桥下死角都使人虽有穿越企图却往往望而却步——这是一种车流占统治地位的、流动的圈地。
穿过一条只容一人通过的狭长通路，来到水边，发现湾的水岸及近海区域停泊了上千只私家游艇。这里为香港赛艇协会，右侧突出海岸为警察俱乐部——一片船的自治领地，全部隶属于私人业主。虽然没有任何标识划定该处的属性，然而它仍然使人感觉到强烈的"排他"感：无形的边界由此而生。

There is one type of border, which is apparently invisible, however indeed exists. It's "invisible" because there is no clear boundary like a "wall" or "fence", its border effect is mainly from the "field influence".
The beach area of Causeway Bay is called "public area", but actually, there are lots of anonymous enclaves. The vehicle road and the lifted highway with extremely fast car flows and big noise make a scary border for the people to pass through: it's a field of cars, a dynamic enclave.
The water front of the Bay is full of thousands of boats, which is tied together to form a "boats' autonomy". They all belong to private owners. There is no sign or fence on the sea, but with such quantity and frequency of "capturing", the boundary is already "there".

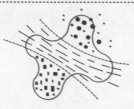

border condition 03
极简城市 （杏花邨）

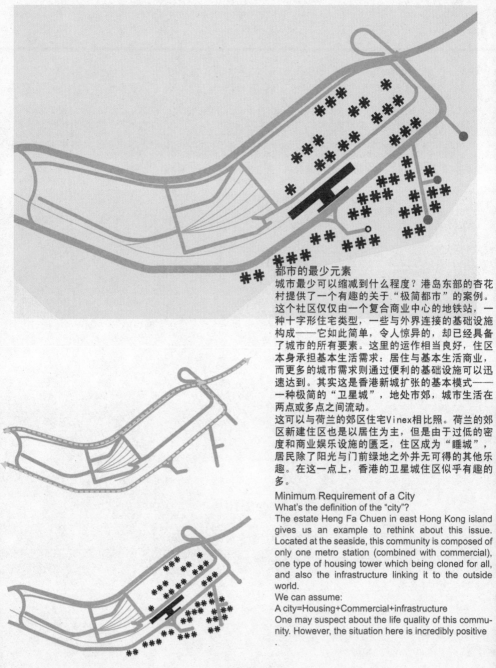

都市的最少元素
城市最少可以缩减到什么程度？港岛东部的杏花村提供了一个有趣的关于"极简都市"的案例。这个社区仅仅由一个复合商业中心的地铁站，一种十字形住宅类型，一些与外界连接的基础设施构成——它如此简单，令人惊异的，却已经具备了城市的所有要素。这里的运作相当良好，住区本身承担基本生活需求：居住与基本生活商业，而更多的城市需求则通过便利的基础设施可以迅速达到。其实这是香港新城扩张的基本模式——一种极简的"卫星城"，地处市郊，城市生活在两点或多点之间流动。
这可以与荷兰的郊区住宅Vinex相比照。荷兰的郊区新建住区也是以居住为主，但是由于过低的密度和商业娱乐设施的匮乏，住区成为"睡城"，居民除了阳光与门前绿地之外并无可得的其他乐趣。在这一点上，香港的卫星城住区似乎有趣的多。

Minimum Requirement of a City
What's the definition of the "city"?
The estate Heng Fa Chuen in east Hong Kong island gives us an example to rethink about this issue. Located at the seaside, this community is composed of only one metro station (combined with commercial), one type of housing tower which being cloned for all, and also the infrastructure linking it to the outside world.
We can assume:
A city=Housing+Commercial+infrastructure
One may suspect about the life quality of this community. However, the situation here is incredibly positive

border condition 04
极简城市（二）杏花邨

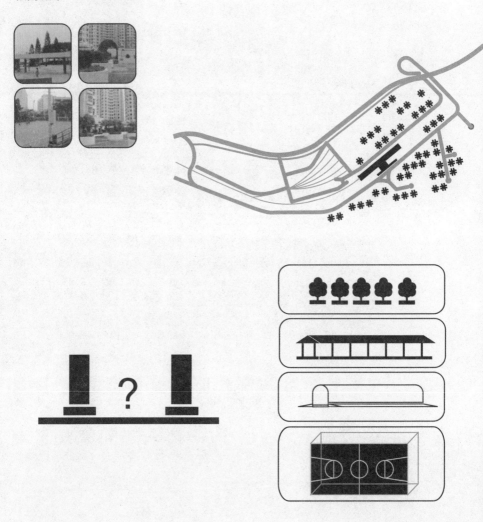

填充的盛宴
住宅的类型是单调的，但是其间的填充物却可以极其丰富。性格各异的公共设施及景观被填充入了单一类型住宅塔楼之间的间隙广场。它实现了某种出乎意料的转化：塔楼的均质性，成为了某种"框架式"的分隔物，围合成一个个城市房间，成为构成这个"微城"的唯一句法；而多样性则由其中的填充程式决定——居住机器的单调性被户外空间的多样性所弥补。

The typology of the residence is monotonous, but the infill could be interesting. There are all kinds of different public facilities implanted into the space in-between the towers. Here, the towers almost become the elements of obstruction, became the only grammar to form the "micro-urban structure", while the diversity is determined by the programs, which could generate at least an inspiration: structural stability combined with programmatic flexibility.

border condition 05
封闭社区 （西湾河）

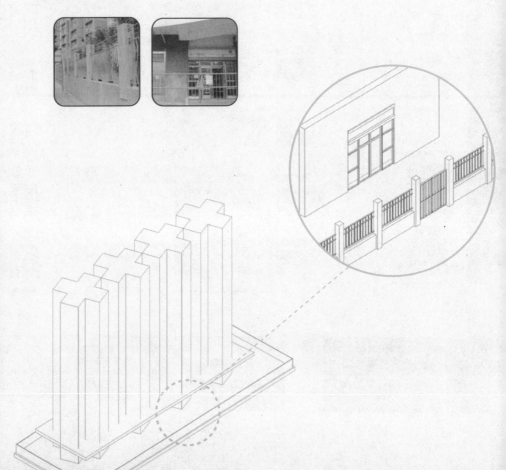

都市的私有瓜分
以安全和私密的名义，香港的新开发住区往往是封闭的，有多重保全系统：外层围墙（或金属樊篱）将周边围合，安全门只有通过密码或者门卡可以开启，加之人工的监管——保安及监视摄影机的监控。院墙是将该住区与外界隔离的实体，而安全门则定义了人群的等级：属于该地的人与不属于该地的外人。
不同程度的围合是一种全城尺度自我孤立的隐喻。我们持续发现，现代城市本身正被划分为各种"特权化"的领域，且这种划分有进一步细化的趋势；更小的权力社区，可见的或不可见的。除了封闭式住区，尚有各种俱乐部，需要会员卡方可进入——看似对所有人开放，实则只为少数人服务。福轲所论述的权力化空间已经在当代都市不可阻挡地成为现实。我们与"公共"似乎越来越远，对于"隔离"的心理需求日益高涨——各种封闭及小团体的自我隔离正是这种心理的实体反映。
是否"不信任感"与"领域占有感"已经成为现代社会的普遍情绪与主流意识？

In the name of security and privacy, the newly built community is gated.
There are several levels of guarantee: the wall, the security gate which could only be opened with pin code, the guard and the CCTV system. The wall is the border which divides the land from the outside, the gate specify again the people, to the level which only belongs to each certain building.
The hierarchy of different levels of isolation is a metaphor for the whole city: the city itself is divided into hierarchical archipelagos. The citizens are pushed more and more into small specific communities, seemingly, we are far and far from "public". The psychological demand for "isolation" is more urgent than the physical protection.
Has "disbelief" become the common emotion to the modern society?

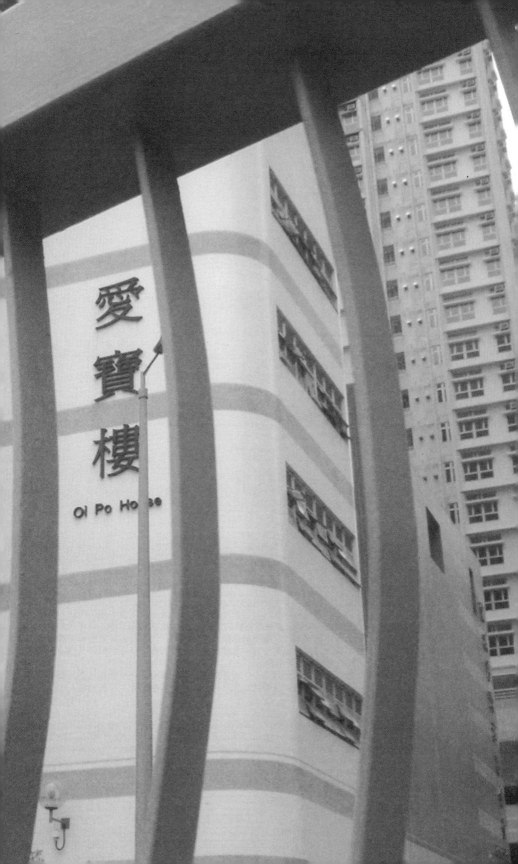

border condition 06
作为激发的墙 （时代广场）

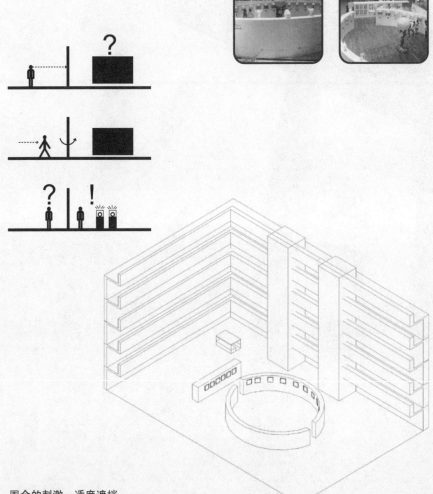

围合的刺激：适度遮挡
时代广场的中庭经常作为艺术家的展场，2008年的一次几米漫画展吸引了青少年的眼球。展场布置在二层的中庭，展览空间由简单的三片轻质隔墙构成：一片直墙，两片弧形墙。其中两片弧墙相对围合，略微错位，留出人群的出入口，展品被悬挂于隔墙内侧。这种围合展览空间并留出些许入口的设计是一个杰作，设计师深知人群心理：从外部看来，参观者可以看到的只是白墙；当你走近开口时，你可以得到艺术品的图景的碎片。
墙激发了人群的对于"里面是什么"的好奇心，而开口使这种好奇心进一步加强，让人忍不住一探究竟——这里存在一个有趣的悖论：通常被认作"拒绝"的遮挡在此处并没有阻挡人们进入，反而刺激其深入的欲望。

There was once an exhibition of Jimmy's comics in the atrium of Time Square. The exhibition space is formed by 3 pieces of walls: one straight, two curved. By cutting a whole circle in the middle and make a slight displacement, the exhibitor creates two small entrances to the paintings hanging inside. The idea of the enclosed exhibition space and the two openings are perfect. From outside, what the visitors could see is only the wall; once you get close to the opening, you will gain the fragment view of the artworks.
The wall makes the people curious about "what's inside"; the opening makes the curiosity even stronger. The wall is not only to "stop" or "divide", it could also lead to the opposite: to attract.

151

border condition 07
24小时超市：昼夜边界

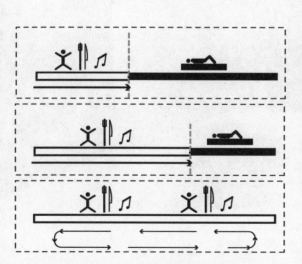

不夜都市的供给
香港密集区的大型超市及便利店大多24小时营业。超市本身的综合化程度非常高，包含生熟食物、日用品、现场加工的果汁及食品、饮品酒类，几乎无所不包。这是都市生活24小时运转的必要补给——即使在凌晨两点，超市内仍然会出现排队购物的状况。如果我们将生活的可激活部分与不可激活部分看作一种边界，那么，在香港，这种边界几乎是消失的——所有时段是激活的，都会生活无限延展。也许就其本身谈论这个问题尚不能使人有深切体会，那么如果我们对比一下欧洲超市：每天晚间8点之前必关门，周日及节假日亦不营业的状况，就可以知道香港超市的优势。欧洲以"人权"为理由不营业——其结果就是绝大多数人的不便利性；而香港的24小时作业也绝非非人性——简单的通过分组轮流当值即可解决。在欧洲却以一种过于简单化的方式来处理——所谓的权益是否真的存在？欧洲都市一片萧条景象，人群的可激活生活周期极其短暂，这是否应当反思？

A day never ends
The 24 hours' supermarket indeed extends the boundary of the day. It's the original power to support the 24 hours' urban life of Hong Kong:
You could get any thing at any time of a day. If the active part and non-active part of life could be considered as a boundary, then this boundary has almost disappeared in Hong Kong.

border condition 08
可变边界：船市 （西贡）

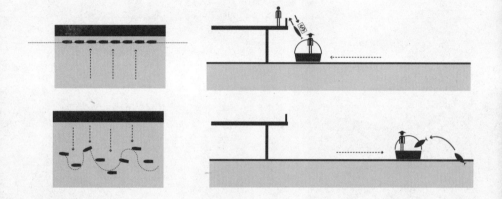

动态边界
边界的改变不仅可以通过时间，也可以通过实体空间上的变化实现。
西贡的渔民靠捕鱼业为生，他们买卖的方式保持了传统的独特性：在每日特定时间将船停在岸和码头的边缘，直接向岸上及桥上的买家出售鱼鲜，他们用滑轮将鱼从船上提升至岸边。数百只船停泊在码头岸边，形成一种"船上的市场"，亦是一种边界。
这是一种作为"媒介"的边界：于渔民和买家之间的一个商品及货币之间的桥梁。这种动态市场的优势很明显：可以将鱼在水中活养，保持吸引力和新鲜；而买家可以顺着岸边走动，观察可供挑选的货品，经历与逛鱼市相同的体验；同时也便于双方侃价。渔其实完全依赖海而生活，在夜晚，渔船离岸分散至海中，继续捕捞以获取第二日的鱼源。此时边界被打散。日复一日，西贡的鱼市边界不断循环着聚合与分散的过程。
西贡的鱼市是一种奇观，一种在现代社会中保留的、真正亚洲式贸易方式的奇观。

Movable Border
Border could be dynamic not only through the dimension of time, but could also be in a actual spatial way. The fish-man who lives on fishing Sai Kung will moor their fish boats by the edge of the bridge and the bank everyday. They then could sell their fish directly to the buyers on the bridge. A pulley is used for getting the fish up. Hundreds of boats stay by the bridge pillars, a "dynamic market" is formed. Simultaneously, a temporary border is also formed there.
This is a border as media, between the sellers and buyers, a connection for the commodity and the money. The advantage of the "dynamic market" is obvious: easy to keep the fish alive in the water, and to be also attractive; the customers could also choose their goods in a marketing alike experience; it is also easy to bargain. The fish-men's life actually fully relies on the sea. When the trade finishes, they ship away from the bank to the sea and get more for selling.
The fish market is a wonder, a wonder of real Asian way of trading, the traditional way of life which still remains in the highly modernized society.

border condition 09
陆上孤岛：联合广场（一） （西九龙）

Island on land
The Union Square which locaated in west Kowloon is urbanized with luxury high rise housing and shopping mall, but surrounded by land which is totally deserted:it is an archipelago on land

The Visitors:
could only access and stay in the shopping mall and terrace but forbidden to the residence

The Residents:
could go to their own property and the shopping mall both, having the view to the sea

The Ring Barrier
The gated communities are another kind of archipelagos, forming a ring barrier to both inside and outside

陆地上的孤岛是指被荒芜土地包围的都市化的区域。它以"高亮点"的方式存在，也因与环境的强烈对比而显得分外突兀。西九龙的西南角有大片"城市白板"，本来拟建于此的西九龙文化娱乐区因为经费来源争议一直悬而未决，而其中的北部地块被地产商捷足先登，开发作为高端豪华住区+高档购物中心，新的国际金融中心也位于此处。因其为联接港岛及机场的地铁的必经之路，因此虽然周边仍然停留在"零度"状态，开发者也丝毫不担心该地的赢利状况。没有建筑的地方，并非"什么都可能"，而是最能赢利的最有可能。

The archipelago on land is the urbanized district within the deserted urban area. It reveals the form of "highlight spot", and becomes subversive, with its strong contrast to its circumstances. Its wider context is originally supposed to be the site for "West Kowloon culture district", which has been delayed long for the investment issue. The government hopes it could be the hotspot to trigger the construction of "west Kowloon cultural district" while the developer only sees the profit. This seems to be the unavoidable destiny of Hong Kong cultural construction: always started and tied with commerce. Although its surroundings are still in the status of "zero ground", the developer has no worry about its future, because it is an important stop on the airport express.
"Where there is nothing, everything is possible" is not accurate, it should be expressed as : the most valuable thing is the most possible one.

border condition 10
陆上孤岛：联合广场（二） （西九龙）

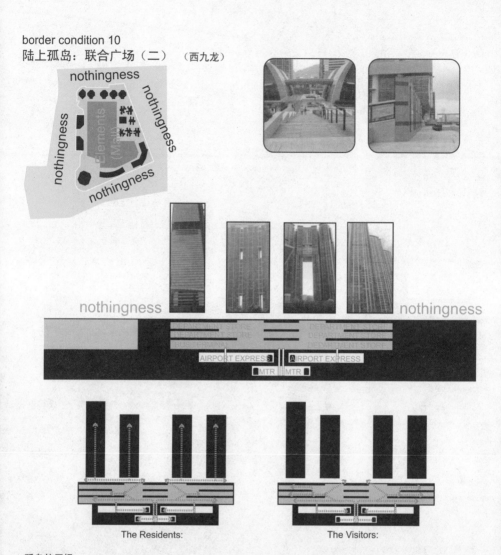

孤岛的层级
孤岛是一个隐喻的定义。在城市尺度上，它地处一片城市的荒原内，四面一无所有，只有基础设施与城市中心相连。被隔绝的状态形成天然的孤岛效应——投资者看重其利益，而政府希望其作为一个先行的"热块"，以驱动处于停滞状态的文娱区这把蒙灰已久的哑枪的扳机，引发周边的连锁反应。这似乎是香港发展任何一个文化区难以摆脱的命运，相似的操作过程：商业先行。这是孤岛的第一层含义——在都市发展的困境下，以一种自我隔离而非"融入"的方式实现都市化。这似乎有悖于城市发展常理，而在此处，远离已经形成的中心、另辟天地、集中于一点的思路似乎比沿中心拓展的方式更有效。
孤岛的第二个层级是联合广场上的高层住区及金融中心各自形成封闭的"圈地"——只对该栋的居民或商务人士开放；公共性仅限于楼下的商业及地铁——甚至连海景和近海空间都是"专属的"。这些元素各自构成了孤岛中的孤岛——永远处于对于人流的过滤状态。

The definition of "archipelago" is a metaphor. It has two layers of meanings:
In the urban scale, it located in a deserted area, the only surrounding is nothingness, its only connection with the city is the infrastructure. In its unique confrontation, the strategy to stay far from the city centre rather than to be close, to isolate itself rather than "melting into" seems to be more rational here than the other way around.
The second layer refers to the fact that the luxury residence and the business skyscrapers forms "enclaves" of their own--only open to their own residents or business men, the public only exist in the shopping mall and the metro----even the sea view and the seaside space is also authorized. These generate the archipelagoes within the archipelago---for ever buffering for the flow----the inherent tortures of this kind of development.

border condition 11
罅隙街市 （中环）

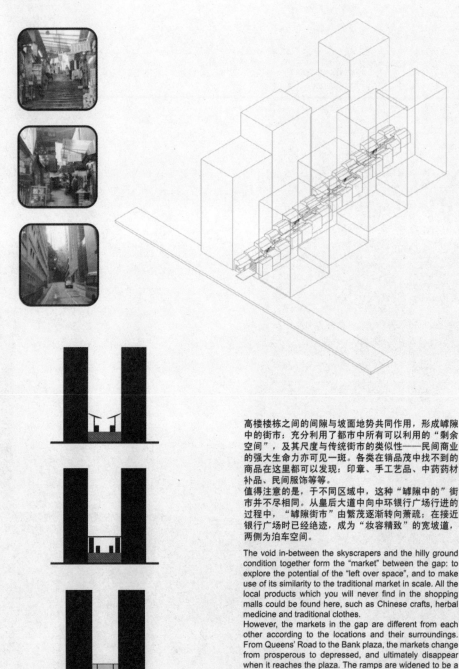

高楼楼栋之间的间隙与坡面地势共同作用，形成罅隙中的街市：充分利用了都市中所有可以利用的"剩余空间"，及其尺度与传统街市的类似性——民间商业的强大生命力亦可可见一斑。各类在销品茂中找不到的商品在这里都可以发现：印章、手工艺品、中药药材补品、民间服饰等等。

值得注意的是，于不同区域中，这种"罅隙中的"街市并不尽相同。从皇后大道中向中环银行广场行进的过程中，"罅隙街市"由繁茂逐渐转向萧疏；在接近银行广场时已经绝迹，成为"妆容精致"的宽坡道，两侧为泊车空间。

The void in-between the skyscrapers and the hilly ground condition together form the "market" between the gap: to explore the potential of the "left over space", and to make use of its similarity to the traditional market in scale. All the local products which you will never find in the shopping malls could be found here, such as Chinese crafts, herbal medicine and traditional clothes.

However, the markets in the gap are different from each other according to the locations and their surroundings. From Queens' Road to the Bank plaza, the markets change from prosperous to depressed, and ultimately disappear when it reaches the plaza. The ramps are widened to be a clean vehicle road with parking lots on both sides.

Border condition 12
边界状态模式研究

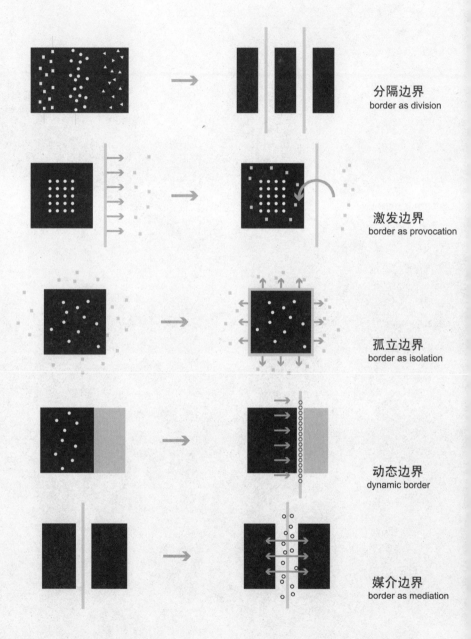

附文：密度相关三例

appendix: three essays about density

8

附文：密度相关三则

appendix: three essays about density

隐形逻辑的效力

——远离思维的积习

一定有人会问：研究香港城市的"隐形逻辑"究竟有什么用？
回答：它其实提供了设计创意的巨大潜能——从多个方面。
那么一定会有人进一步质疑：既然这些"逻辑"已经是城市中的存在，那么，我们，作为城市的缔造者，即使明了并继续套用这些逻辑，是否仅仅是对于既有现象的简单复制，或者是对既存城市语法的重复滥用而已？如何能在"现实"中发现新的可能？
答案是：隐形逻辑的创意潜力及完全不同于以往的都市更新的动力，关键在于对于既有逻辑的"转用"和"放大"。
日本建筑人士贝岛桃代及冢本由晴的城市研究文本《东京制造》对于东京"杂交现象"的研究，在暴露都市现象方面视点敏锐、观念犀利，可是它并没有提出这些现象的研究对于城市的再发展有何启示或指引作用。它专注于客观现实的描述而缺乏理论的建立，它注重再现而非创造。虽有评述，仍然是对于混杂现象本身的问题，诸如伦理、程式等的评价，并没有更多的对于城市的进一步主观介入。
"隐形逻辑"对于现象揭露之后仍然有所评述，其意旨在于为设计及城市重整提供批判式理论依据。我们不妨从新梳理一下"隐形逻辑"的主题，明确其对于设计中概念创新的提示与意识构建的可能作用。我们仍然按照研究中案例的顺序依次进行剖析：

标识系列

1) 被广告包覆的裙房立面——提示了建筑内外的再次分裂：程式、立面与广告。我们是否可以从中间抽取立面层？那么这个结构就将变成program——advertisement的形式；或者我们是否可以对这层关系进行重组，由内至外依次为程式——广告——立面，前提为立面为透明或部分透明。这时广告层则成为中间层，它可以时常更换，更具灵活性，适合更小的分割，并且可以成为立面展示的一部分（建成的例子是北京建外soho的立面的内置广告）。

2) "牛扒店"与"小食店"的标识策略是"密集化"策略、"无所不在"策略及"高频率"策略——如果将其原则放大至极大，将引起质变。

3)"欢迎楼梯"表述了建筑的"仪式性"在传达空间心理方面的作用——通过空间的处理传达建筑的某种"态度"。建筑的客体性被消解,成为具有感性的主体。空间如果具有某种可感知的情绪,那么这已经是一种设计上的突破。
4)"最大化吸收的东角中心"证明了建筑构件与通道开合对于人流控制的作用。我们可以继续发挥建筑的"控制"作用么?还是相反力图将"控制"减少至最小?更有趣的是"兔女郎"提示了一种新的控制途径——"软控制"。探讨"控制"与"放纵"的多种可能形式已经可以单独作为一个研究课题。
5)"地下迷宫"与"二层书店"的繁华都有些出人意料,正是这种"出人意料"证明了在人们习见中存在已久的、对于"消极空间"的定义已经不再牢靠。如果"消极"不一定消极,我们似乎可以重新开发"不利地带"。

品牌系列
1)"LV"系列提示了在变化中保持统一的微妙性;
2)"Fendi"系列表明装饰、标识和构造可以是一个统一的整体;
3)"Dior"强调了"整体氛围"营造的手法,甚至通过反常逻辑的反响操作。
4)"Calvin Klein"的摩天广告楼是一种概念嫁接的创举:广告牌与办公楼的杂交。那么进一步转化的话,我们甚至可以把办公楼做成广告牌的形式——例如MVRDV为西班牙APAVISA公司设计的办公总部。

密度相关主题
这一章节是对"高密度"下空间特殊处理与特殊城市空间的总结。
"杂交"
1)"堆积":三种空间结构组织的叠加——足以对抗简化至极的"标准层无限复制"模式。如MVRDV在汉诺威世博会荷兰馆的设计中,采用了相似的逻辑:一个建筑程式的"三明治",将完全异质的程式及体验的差异在垂直的方向累积——它是异域的、陌生的、无前例的。
2) "侵染":分属于不同精神领域的建筑之间的相互渗透——以程式的形式。那么,我们同样可以设想空间的"等量交换"——程式的交融在形式上可以进行表达。
3)"城市进化的断片标本":在城市的保护与更新中,可以考虑采用分区的方式,使建筑成为都市进程的"标本"甚至"化石"。建筑的纵向层次并置可以类比于地层不同地质的积淀。新旧共存的方式下,对于旧事物的反响并非都指向"不和谐",而是更复杂有趣的"城市层级"。

4)"中心秀场"指引了通过集中化都市象征式组件及控制人流方式达到"都市图景"与"心理"的构建。
5)"地铁名品综合体":两种通常冲突的程式紧密、平滑地"无缝拼接"在一处。我们不禁重新提问:"什么是不可能的?什么是常理?"——设计中的"常识"往往是最不可靠的。

"共生"
兰桂坊酒吧在街区内的六种共存方式,为在逼仄条件下争取生存几率提供了样板——我们可以进一步想象一个所有空间都以非常规的方式组合在一起的建筑么?组件本身都是非均质的单元——我们可以达到单栋建筑内所有生活空间的异化——仍然是"放大"和转化的方式,现象被转译为构造"句法",然后放大运用于全体——奇观出现,生活改变。

垂直都市主义
1)"山地摩天楼":在峭壁上建造摩天楼本身就是一个创举,无论在概念层面或技术层面。如何将地形的不利条件转化为空间丰富性的有利条件?都市的向上伸展及水平蔓延可以不再以"层"为单位,而是以"楼栋"为单位——3D都市的可能。
2)"超薄":在竖直方向无限延伸和水平方向无限缩减的悖谬统一体,它本身就是一种"超常规"都市景观。
3)"悬挑":在水平方向上争取更多空间,又不超出规范规定的边界要求的唯一选择,可试用范围很广(例如MVRDV在阿姆斯特丹的WOZOCO老人公寓就这种手段的最大化)。
4)"高密度下的空白":在高密度下仍然得以存在的空白空间的生命力在自于其所具有的强大公众影响力及对于过于饱和的城市的"呼吸"和"缓冲"作用,可以作为我们设计高低密度混合模式,以及设定密度原则的依据。
5)"高层住宅的极端类型":基于高密度的类型学实验,为"密度"与"形式"的可能性设立了原型和参照系。

暧昧的公共性
1)"时代广场的公共性争议":商业的私有属性、盈利性与公共空间的公众属性、非盈利性之间的对抗与平衡。使我们思考现代商业社会挤压之下,公共空间如何求得生存之地?公众可以通过何种手段维护自己的公共空间?
2)"西湾河滨海社区公园":一种同时为公共及社区共有的公共空间模式,一种系统化的组织方法,一种以"情节"激发游历乐趣的环境空间,一种以小尺度营造大格局的手段。

3)"鸟亭":公共设施由广普性向特殊性的转化,提示了"使用"对于公共空间属性的"再定义"作用。这与"少规划"理论很巧合地契合了:是否我们今天的城市状态更期待一种无指向的设计,建筑师应当尽量减少主观介入,而将空间的最终定义权力留给使用者?
4)"基于时间的公共性":在有限的空间资源条件下,非公共空间可以在特殊时段通过分时段转变使用者来提供灵活的公共性——利用时间向度解决空间短缺问题。
5)"商场内的座椅":证明了基于"人本主义"的经营策略才是真正的对于商家和消费者双方都有利的双赢策略。商家不应以短视的小利而忽视了公众需求,放弃长远利益。任何对于公众尊重的细节都可能导致空间属性的完全转变。

非正式
1)"阳台的重构":居民对于城市的自发性改造,体现了民间对于更多空间的诉求,与都市"颜面"的不可预料因素。
2)"程式化的立面":如果能从立面阅读出其中发生的内容,那么这种立面已经突破了"立面"的传统定义。
3)"桥下空间":对于大型城市固体物件所遗留下的大量剩余空间的灵活利用,可以重新激活这些所谓的"灰色地带"。
4)"附着式商店":可消失与再生的街道商业,"街市"的生灭由个体的折叠与展开而决定——街道竟然也可以"折叠"。
5)"折筑"与"伞筑":对于日常非建筑的小物件的非常规再开发:使其建筑化,具有建筑的基本属性,并保留最大的灵活性。

效率最大化
1)最高效地铁:高密度城市的发展必须依赖以地铁和轻轨等快捷高效运输系统为主的公共交通系统,而精确的管理和全民对于社会秩序的自觉遵守才能使系统运转不停。在香港更重要的突破在于地面高密度对于地下延伸的潜力的史无前例挖掘:地下的容纳能力到底有多少?
2)无水的威尼斯:以"分流"的方式为高密度城市多年面对的"瓶颈"提供了一种可行的解决方式,解放了地面效率,为市民的公共生活创造了一个完全人工化的、连续的室内环境,并且意外地收获了城市孤岛之间的联系。

边界条件
现代都市中无所不在的边界,无论可见与不可见,都在左右着我们的生活。哪些边界是积极的,哪些是消极的?我们既可以利用边界的积极效应来达到激发、保护、展示、区分的种种目的,也可以尝试避免不必要的边界。而对于边界的强化、表现或者拆解都可能成为设计中创意的契机。

解析"低密度之梦"

——给欧洲都市泼点冷水

"高密度的优势"的提法让人觉得荒谬——因"高密度"在人们的想象图景中一直与过度拥挤、资源匮乏、喧嚣混乱联系。但是在这个"偏见"与"成见"无所不在的时代,经济学家、逻辑学家、哲学家及所谓"理性选择的倡导者"已经一而再的让事实向人们证明:所谓"经验"往往比"非理性"更加"不可信"。对于高密度的具体表现`,我们有必要重新审视。我仍然认为低密度不但不是所谓的"理想城市",反而问题深重。

"高密度的优势的成立"必定有"低密度的劣势"作为参照。那么我们就以被称作"各国人士所憧憬的理想城市:欧洲低密度城市"进行剖析对比,看看低密度的生活究竟如何。欧洲一直以"美好生活"的代表自居,成为"宜居城市"典范备受推崇,历年的世界最宜居都市综合排行榜中,欧洲的几个国家,如瑞士、挪威、瑞典都高居榜首:"阳光,蓝天,清新的空气,无比悠闲的生活,从摇篮到坟墓的福利保障……"

"针对城市社会政治环境、经济环境、社会文化环境、医疗与健康、教育、交通与公共服务、休闲娱乐活动、消费产品、住房及自然环境9个类别的39个指标,Mercer对全球城市的生活品质也作出单项及综合排名。2003年调查显示,卢森堡为全球最安全的城市;综合生活质量最高的是苏黎士,其次是加拿大的温哥华,随后是维也纳。"

低密度真的如此美好么?对于欧洲的印象会因为你与其相处的时间不同而得到迥异的结论:如果你仅仅在欧洲做一周至半月的短途旅行,你可能觉得它是天堂,但是如果你长年居住于此,你可能会觉得它是荒漠,甚至是因"荒凉"而生的地狱。实际上,欧洲的所谓都市优势似乎越来越只能通过"世界的观光机器"来证明。这种说法可能与中国民众一直以来对于欧洲的美好想象相去甚远,让人觉得难以接受。

从城市角度看,欧洲的低密度一直存在深度隐忧,"低密度"并不是造成这种危机的原因,而是一种表象。这种低密度,更导致城市呈现出的一种整体低迷与萧条——从经济至生活的方方面面,根源在于其体制固有的、隐藏的诸多弊端,如今已经一一显现,成为制约欧洲发展的巨大瓶颈与沉重负担;而让人费解的是,欧洲人自己对于这些危机尚处于一种普遍的无意识状态——即使政府和部分民众意识到了也无力改变。

我以下所描述的欧洲，也许是你从来不曾想象过的，也是从未有人深入讨论过的欧洲。然而这些的确都是现实——揭示被人们所忽略的、却重要的现实，正是研究的客观性所在。

作为西方文明的发源地和近现代在科技及文化上的长期领先，欧洲民众在积累财富的同时也积累了文化上的自我优越感。与他们实际接触之后会发现，欧洲民众固有的对于其自身文化与体制的高傲态度，已经导致任何对其质疑或者异议成为不可能。对于其城市元素的单调性，因为一直以来单向度的判断标准使然，他们可以视而不见。

因此当库哈斯在对世界各地的社会与城市状况以冷静而敏锐的视点进行观察和剖析之后，以较客观地态度将"外面的世界"呈现给欧洲时，欧洲人的态度基本是抵触的，不屑的——无论是对于纽约还是对于亚洲的新兴城市的都市化。欧洲式的城市理论的批判性从来都是建立在一套欧洲传统话语的正当性之上，对于梦呓般的所谓"完美"和"和谐"的追求。而从未有人思考，他们所预设的完美是否真的是"完美"的？和谐背后是否隐藏着深层次的不和谐？

"不想赢利"的商业

欧洲民众与学界一直认为自己拥有悠久的历史，完整而优雅的城市景观，悠闲的生活及洁净的环境，但是，他们唯一忽略的，也是最重要的一点是：都市不仅仅是一个精致的外壳，它更多的是关于生活的具体细节，它的本质是"人"。

欧洲都市的街道，入夜时分即空无一人、一片萧索，所有的商店全部关门歇业。如果你在街道上徜徉，你就如同一个孤魂野鬼，你会有这样的错觉——这个城市是不是死的？它一直陷于这样一种悖论之中：如此的精致和规整，却如此死气沉沉。

据报道："出于保护劳工权益的初衷，战后德国一直严格规定普通商店的营业时间：周日必须关门，平时也只能营业到晚上8点。因此，德国在晚上8点以后大街上就一片静寂。随着时代的变化，这条僵化的规定遭到各方面的猛烈批评。"

欧洲的商业街（除了最中心城市的旅游热点区，如布鲁塞尔的老城中心广场），所有商店在晚上6、7点钟即歇业，并且周六、周日也大多不营业。

荷兰鹿特丹的商业步行街lijnbaan，工作日白天营业，周六开放半日，周日完全歇业，这是法律所规定的工作时间——以"保障人权"的名义。于是举世罕见的奇观出现：当上班族下班、学生下学之后，想满足一点基本的购物需求，竟然不可能的，所有的商业活动都已停止。营业者遵循一个简单逻辑：你们上班，我们也上班；你们下班，我们也需要下班回家——以绝大多数人的不便利性，以整个经济的景气的丧失为代价，维护极少数人所谓的绝对"平等"，这是真正的平等还是仅仅是一种字面上的平等，甚至是一种更深度的不平等？带着对于"绝对平等"的热情，欧洲的民主有时候反而从事了一种对于民生的讨伐。

如果稍具变通意识的话，这个问题并非不可调和的矛盾——如果政府一定要保证人人每天8小时的工作制，也完全可以将其营业时间推后，晚开业、晚歇业——因为稍有常识的人都知道，商业在夜间才是最大的需求时段。而在西欧，政策竟然僵化至连一点基本灵活调控的意识都不具有。这条法律从1958年就已经开始实行，至今长达50年时间。

全球通用的商业基本原则在这里不适用。于是，欧洲的购物成为一种只有平时赋闲在家的退休老人才能进行的活动。在欧洲生存，你必须学会一种能力：囤积货品的能力，储存粮食的能力——因为大多数人可以获得的购物时间并不多，这在今天全球经济时代听起来似乎是天方夜谭，欠发达地区似乎都没有这种现象，而号称高度发达的西欧，世界最富裕地区的人群竟然要为有限的购物时间困扰，这不能不说是一种讽刺。

所以在周末的超市里面你可以看到市民整车购买包装食品及蔬果，然后储存于冰箱。到了法定假日，更是要提前作好准备：如果你节前没有囤积够足够的食品，你可能要挨饿——因为法定假日商店绝对是不开门的。这和亚洲无比兴旺的节日经济完全大相径庭。

在亚洲生活的时候，你可能会抱怨节假日大街上人太多；到了欧洲，你会开始怀念"亚洲岁月"，因为毕竟"人太多"的问题，比起周末"什么都买不到，什么都不能买"，或者"饿着肚子过节"之类的问题，已经算不上问题。香港一位瑞士建筑师的妻子是菲律宾人，她随他先生去欧洲度假（请注意，瑞士是全球综合生活质量评比中级别最高的国家之一）她不到一个星期就回到香港了，原因是"在那里的无聊和不便使她几乎疯掉"。在欧洲人眼中经济欠发达地区居民来到发达国家，竟然无法忍受那里的生活，这结果多少有点出人意料，却在情理之中。

"发达"的服务业？

欧洲一直以服务业发达著称，服务业是都市生活质量的一个重要指标。服务业发达应当等同于"受服务的人群生活便利，舒适度高"，然而事实果真如此么？

在欧洲，你会发现一个很奇异的现象，就是欧洲人自己修自行车，自己在家做饭而不怎么去外面餐馆，甚至自己给自己理发。这些并不是他们的"闲情逸趣"，因为这更多的是出于生活切实需求。

在欧洲如果去车行修单车，例如补胎、换车闸之类的小问题，其花费等同于重新买一辆二手单车。而且修车还需要预约，预约了还不一定能约到，他会告诉你"很忙，至少要等到一个星期以后才能取"，所以大部分欧洲人自己在家修自行车——他们从小就精通此道。人们也自己理发。MVRDV有一位建筑师，已经做到项目经理的位置，一天他理了发，我问他，"发型不错，在哪里理的？"——回答是"自己理的"他说"理一次发要50欧元，太贵了，我还是自己弄好了。"

在香港，一个普通工薪阶层也可以每天下一两次"馆子"。在欧洲这是不可能的，对他们来说太奢侈了！欧洲在餐馆，一顿正餐至少20-40欧元，看上去似乎并不多，但是如果对照一下他们的收入，大部人每月不到2000欧元（税后）的收入，如果每天"下两次馆子"，那么他们基本上不用再进行其他任何日常生活开支了。所以，欧洲人平时大部分时间自己在家做饭；或者做简单的三明治，或者煮意大利面。

亚洲都市的出租车满街跑，有需求时招手即可得的服务在欧洲是不可能的。荷兰的出租车需要打电话提前预约，大街上几乎是很难看到出租车的影子，有急事要打车，很难。这是否也是"服务业发达的体现？"

最后我们总结出结论：原来，欧洲的"服务业发达"是指"从事服务业的人群收入高"——该产业创造的收益高。因此以上命题的直接推论就是：为获得某种服务，大众的投入成本高；间接结论就是：大众普遍享受不起日常服务。

过度福利的隐忧

近年来中国的学术界有一种倾向，知识分子和经济学家在媒体上纷纷提倡中国应当效仿欧洲的福利模式。基于改善民生的善意初衷本无可厚非，却因为过于主观而忽略了对于实效的真实研究。实际上，作为高福利制度的创造者的欧洲，目前正在福利制度的深重负担之下，"有苦难言"。

中国广大民众也一直凭各种传言和报道，认为欧洲福利制度从家庭补贴、免费教育、失业救济、医疗保险、养老金组成了一个天衣无缝的福利网，把欧洲人（特别是西欧、北欧），从"摇篮到坟墓"都照顾得体贴入微。殊不知今天，这些"幼有所依、老有所养"的乌托邦在风光外衣之下，早已岌岌可危了。

"自2000年以来，欧元区年均经济增长率只有1%，失业率居高不下（最高达到8.6%），欧洲国家原先引以为豪的福利制度已成为经济发展的重负。在德国，失业工人可以得到原工资67%至53%不等的失业救济，加上住房、小孩抚养等补助以及免交税款，一些失业工人的社会福利待遇甚至超过低收入者的收入，致使一部分失业者宁可在家闲着，也不愿从事低收入的工作。而近年媒体报道的一系列合法却极不合理的领取社会救济的案例，更是引起全社会哗然。一个社会救济金领取者长年居住在美国佛罗里达州风景优美的海边，优哉游哉，而德国每个月都给他汇去衣食无忧的救济金。更有甚者，一个参保者服用伟哥以提高个人生活质量，其昂贵的药费经法院判决由保险公司承担。"

——南方网讯

欧洲人在温床中丧失了昔年那种积极进取的锐气，不断增长的长期失业人群已经破坏了社会的完整性，以至于欧洲人现在有这种心态：既然不工作都可以获得收入，为什么还要那么卖力付出呢？在德国450万的失业人口中，超过50%已失业一年以上，长期与技能的疏远使他们的就业前景更为黯淡，更依赖于社会福利。但即使愿意工作，失业者重新找到工作的机会也正在逐日降低，这不仅因为他们的技能变得生疏落伍，而且是因为企业在转移工厂后，所提供的就业岗位正不断减少。

欧洲最大的经济体德国的问题尚且如此严重，可以想象，整个欧洲的高福利制度遭遇到的冲击有多大。这些国家都面临着相同的困难，其中，出生率下降、寿命延长、人口老龄化最具普遍性。20世纪60年代中期，西欧的失业率只有美国的一半，到2000年，欧盟国家平均失业率为10%，失业人口约为1600万。而美国近年的失业率则低于5%。以上种种数据都显示了如今的欧洲创造福利的人不断减少，享受福利的人却不断增加。

"在瑞典政府把税收大多投入到社会福利上后，瑞典失业者从政府领到的失业津贴，55岁以下的人可连续领300天，55岁以上的人则能领取450天，而金额相当于本人工资的90%左右，所以在一些瑞典人看来，失业只不过是一段带薪休假而已，这也使很多人不愿再去寻找新的工作。"

对于福利制度的另一个误解是认为他们是所谓的"高收入者高税收，低收入者低税收"——因此是相对公平的保障低收入阶层的利益。高收入者高税收是事实，但是低收入者却未必是"低税收"。荷兰的个人所得税起税标准是33.3%，这意味着只要开始工作，就至少要交33.3%的税收，而这个是针对所谓"低收入者的低税标准"。如果你的薪水一个月是2000欧元，那么你实际所得只有1200欧元——三分之一以上要以税收形式在拿到工资之前就已经被直接扣除了。请注意：这还仅仅是起税标准。随着工资的增长，税也快速增加，有的甚至高达60%至70%——一个不工作领取失业救济的人一个月尚且有620欧元的补助，而每天辛苦工作的人的工资却与他们相差无几。福利不是天上掉下来的馅饼，而是"羊毛出在羊身上"，靠工作人群自己平时递交的税收来积累。并且，实际上，所谓的养老保险，免费教育，医疗等等的支出与你一生所交的税相比，几乎是九牛一毛。那么大部分的税收哪里去了？就是流入了一大群明明具有工作能力，却什么也不做，在家坐领失业救济的人的口袋里。

"瑞典北欧斯安银行的经济总监克拉斯.埃克隆德曾不无抱怨地说，瑞典的税收高到了令人恐怖的程度，他每个月的收入有近60%都要用于交税，这几乎使他丧失了多赚些钱的积极性，而长此以往的结果就是，像他这样的高收入人群会因为高税收而变得不思进取，而那些失业者则可靠政府的福利衣食无忧，也同样会失去工作的愿望，最终导致了国家的低效率。"

这种不加区分的"统统福利"在本质上和中国在过去计划经济时代曾经令国人痛苦良久的"大锅饭"没有区别。我们虽然离那段岁月已经遥远，但是不要忘记那种一锅端平的"平均主义"给我们国民经济所带来的严重制约和挫伤。欧洲这种过度福利的体制，实际上正在促使欧洲成为一种"养懒人"的社会。用一句上一辈人比较熟悉，而我们和我们这代以后的人都不曾深刻体会过的30年前的流行语来说就是："干多干少都一样，干好干坏都一样。"——既然付出劳动与不付出劳动结果都一样，那么谁还愿意去工作呢？

欧洲各国政府对于福利制度所造成的弊端早就头痛已久，而呼吁改革的声音也越来越大。但是往往到了关键时刻就遭遇挫折，主要是因为福利的既得利益者（那些习惯了"衣来张口，饭来伸手"的人）不愿意改变他们现在不劳而获的生存状态，且这部分人数量众多。而欧洲政党在执政期间对于平衡"民意"方面谨小慎微，为避免失去支持率，避免广泛的抗议，而往往很难有大的举措。所以"高福利制度"的包袱背了很多年，越来越沉重。无奈之下，欧洲各国政府只有通过提高税率以支撑日益增长的支出。但这样又引发了一个新的危机：企业成本不断高涨，加上欧洲在全世界位列第一的劳动工资，企业无奈之下只好纷纷离开本土去劳动力成本更低的国家设厂，造成国内失业率进一步飙升。我们常常论及中国经济问题是"经济过热，增长过快，结构不合理"，那么，经济长期停滞不前难道就不是问题了么？

看似优厚的社会福利，从实效上判断，造成社会资源的浪费。瑞典号称世界福利国家的典范，拥有最为各国人所羡慕的社会保障体系。然而近年来，这一福利制度对瑞典发展的利弊，也成为该国大选的主要争论焦点。最终，宣布将对福利制度进行改革的"右派"政党联盟获得了更多选民的支持。

欧洲各国政府已经显示了对于福利制度的调整所下的决心。可以想象，决断的、甚至是冷酷无情的福利改革必将给享受惯了的欧洲人带来强烈的刺痛，甚至可能会激起广泛的抗议，但却是不得不进行的过程。若及时加以变革，经济增长率和就业率还有望重上轨道；如果继续为了面子而不切实际，后果将是无法填补的经济黑洞。

都市图景

我们呈现了欧洲不景气的商业，昂贵而难以承受的服务业，以及福利制度对于整个欧洲的"锈蚀"效应，那么，这些内容究竟与"城市"有何关联？

在前文已经指出，都市并非仅指组成都市的客观物质实体，如建筑、街道和广场，它更多的关乎真正的生活质量。对于都市实体面貌的体验，仅仅是都市体验的一部分，而生活质量的体验，却恰恰是由上文讨论的内容所决定。

欧洲现在处于这样一种状态：在华丽的城市外观掩盖之下，都市生活处于普遍的匮乏和萧条之中。人们日常生活处处受到制约，单一而僵化，便利性无从提起；都市之间缺乏有效联系，达到任何一点都需要费尽心力；一种掠夺式的高税收政策将民众生产进取心彻底清除，经济萧条，少部分的城市密集区域之外是大片的荒地。都市本身已经沦为一种纯粹的舞台布景，一种并不照顾切实需求的展场，况且大部分产品都是来自于历史的遗留物，新的建造只在极小的范围内进行，动作缓慢，甚至已经停滞。对于"过去"的过度沉醉，对于既有体制的过度依赖，使任何社会的更新都难以开展。城市未入夜之前已经呈现出睡意，使人难免怀疑其是否尚且可以称为"都市"？

我们谈论欧洲都市的萧条，因为其正可以作为高密度的反向参照物。只有在这种对比之下，高密度中的效率、集中、便利与丰富性才能够真正凸现；也正是于此，"密度都市学"的启示性才可能更显出其价值。当然，如果你对于生活的唯一需求就是"空气清新、安静、居住宽敞"而可以忽略其他一切的话，那么你仍然会更钟情于欧洲的模式。

但是，有必要指出的一个很有趣的现象是：如果对比90年代以前与最近10年去欧洲留学或工作的两代人对于欧洲的看法，会得到完全相反的印象，上一代人几乎是异口同声的羡慕，而我们这代人则更多的是震惊——不是震惊其优质，而是震惊其"反面效应"。所以，他们最终大多选择归国——也许，这是时代的参照系发生了改变的缘故。

地狱的出路

面对桎梏：妥协还是突围？

库哈斯在《S, M, L, XL》的非正文部分有一段非常精妙的关于"苦海"的论述："生活的地狱并不是在将来的某时，如果有，那它早已存在了——因为我们所处的地狱往往是我们自己一手制造的。想要逃离苦海有两种可能：1. 接受地狱并成为其一部分，你再也感觉不到地狱及痛苦；2. 以持续的警醒、怀疑和冒险的欲求，找寻哪里是地狱之中非地狱的部分，将其隔离，赋予其生存空间。"

我认为这段话基本上表达了面对城市与建筑问题中的限制条件的两种态度与处理方式。香港建筑师采用的是前一种，而荷兰建筑师则属于后者。虽然欧洲城市在诸多方面存在弊端，但是欧洲建筑师在设计中的投入、创意和执着是遥遥领先于香港的，这一点毋庸置疑。

1. 将"镣铐"转变为"触媒"

香港建筑师在设计中的"无作为"现象，首先体现在他们或者以城市开发为地产商所主导，或以市民对于建筑投入的有限预算难以做出有创意的设计为由，主动放弃设计。严迅奇与廖维武在关于香港公共建筑竞赛的对话中提到："香港目前还没有这种把建筑视作文化实体的能量，香港总是偏重于强调效率、技术力量和实用性，而创造性和文化价值也就摆在了较低的位置上。"

其实，限制条件阻滞创意的命题本身就不成立。我们仍然以欧洲为参照。在欧洲，业主对于建筑的要求同样苛刻，甚至更为严苛。荷兰事务所MVRDV能将业主的条件作为形式生成的重要依据，每一次设计都会仔细考虑。各种限制，例如场地、经费、光照、流线，对于他们来说不仅不是制约，而且往往是做出非常规的、有趣的设计的促成条件。最经典的将限制转变为创意的例子就是他们早年在阿姆斯特丹wozoco老年公寓中的建筑处理：在有限的基地地块内，业主要求置入100个单元的住宅。而根据当地的建筑法规，要满足容积率要求则建筑会对周边的建筑造成过多遮挡；按照限高要求只能置入87间住宅；如果将这13个住宅置于周边又会占用社区本身的公共空间：建筑似乎已经走入死角，被无可解决的矛盾所牵制。MVRDV并没有因此放弃，是否仍然可能找出某种既不对周边造成阻碍，又能达到容积率要求的解决方式？他们想到了一种非常规的手段：悬挑。剩余的13个住宅被以出挑的方式置于建筑主体之上。这样建筑既不会增加额外楼层，也不会影响地面使用，并且更重要的是——创造了惊人的建筑奇观。这至少说明，限制条件并不一定是"难以创新"的完美借口，如果具有探索精神和灵活应对问题的思路，"限制"甚至可以成为创意的激发剂。

2. "熟悉"元素的陌生化

香港建筑界普遍存在一种潜意识：好的设计一定是昂贵的设计。而这完全是一种对于设计的误解。当我们注视着妹岛和世的作品时，它们往往在形式上极为简单——或者是最常见的方盒子，或者是极其简单的抽象形体，但是却往往传达出一种非常"陌生"的意向，与我们所熟知的、备受诟病的现代主义国际风格完全不同的观感——而这尚且仅仅是对于形式本身的讨论，尚且不涉及设计更有趣的program部分。MVRDV的建筑也呈现出相同的特质：以人们最常见的、最熟知的元素构造完全迥异的建筑。

香港建筑师担忧的问题的核心是：建筑的创新会导致造价的升高，而这是精打细算的业主和公众所不能接受的。而建筑造价的飙升主要是由两个因素主导：非规则的形体或者大量扭动的曲面，导致需要繁复的受力计算及难以标准化的构件生产——每一个构件都是不同的，成本自然提高，例如Zaha的很多设计都有这个特征；另一种因素是昂贵的材料。建筑师提到创新，往往联系到夸张的形式或者所谓震撼力。这正是对于建筑本质的简单化理解所致。图像和消费的确助长了这个时代的浮躁情绪，但是设计的创新仍然可以有更深层次的含义。它更多的是在程式上，在对于生活方式的作用上的一系列根本性的创意或者对于传统的颠覆，而这种创意完全可以通过对于平实的形式的"陌生化"处理，对于将不同的概念重新组合（例如思考将办公与娱乐的结合）来达到，结果并不比"普通建筑"造价提高很多。

3. 最大化差异的混合

面对密度的限制，面对建筑法规的限制，香港的住宅永远只能采用从户型到楼层千篇一律无穷克隆的结果。有谁想过通过简单的，将不同户型拆散、叠加、重组，就可以产生出完全异质的住宅类型？

荷兰近年来的城市扩张中，出现了叫做Vinex的新建住宅区。1993年末，住房部长选取了荷兰一些半郊区、卫星式住区的地址。拟在2005年之前新建60万个新住宅，并且规划了7个主要城市圈的30万新住宅量。Vinex是一种同时以个人化和全球化定义自我，标示社会变化的宣言。

Ypenburg是个典型的Vniex住区。这与香港的新城发展状况情况类似，为了减轻市区发展的压力，降低中心区的过高密度，分散人口，香港政府自1973年起开展新市镇发展计划，至今开发了九个新市镇，分为三代。第一代为70年代初动工的荃湾、沙田和屯门新镇；第二代为70年代末的大埔、粉岭、上水及元朗；最近的三个新市镇——将军澳、天水围及东涌发展工程由80至90年代开始发展。现今整体人口320万，预计到2013年增至340万。这些新镇是作为"小型城市"来设计的，在功能上基本满足了自足的要求，也有相应的基础设施与中心连接，可是仍然避免不了在建筑类型上的单一性和简单化处理。

MVRDV为海牙的Ypenburg所做的规划是这类郊区住宅中的成功案例。将整个地块以一定的网格划分为面积相对平均的区域，而每一片区域则由不同的事务所，按照不同的逻辑进行设计，创造其独特的场所个性。基本逻辑是"均质的结构，而差异最大化的内部填充物"。形成了合院型、联排型、船屋型、随机型等多样化的，基于地块的社区组团。

MVRDV在其负责的Waterlijk住区之间留出的不同尺度的"中间空间"，强化了这种空间的多样性。它仍具有私人的边界，但是公共空间如绿地、铺地、小品和道路之间互相渗透。所以在感觉上，私有区域只是公共区域的一个个外花园。建筑本身仍然是模数化的，基本单元也是标准化的，而其中的创意与差异体现在对于不同单元的组合方式以及建筑中间空间的创造上。而在建筑的处理上，也体现了建筑师对于公私的分界以一种微妙的方式实现：优雅而含蓄。

其中一块Maccreanor Lavington设计的L型住宅，数个相对的L围合成传统的庭院。它并没有创造新的户型，而是将灵活性通过对于使用的非限定性来表达。它临街的一面采用高阳台和高窗，在视平线以上，而面向较私密内庭院的一面采用低窗。在Ypenburg Singels区的住宅，建筑师有意突出了其城市性，他们研究了不同类型的组成方式，不同剖面标高的住宅被并置在一起，甚至屋顶起坡的角度和方向也经过精心设计。

4. 放弃对于"成功原型"的无止境复制

香港电影《无间道》获得成功之后，其后三年的警匪片主题立刻全面被"卧底潮"占领，从《黑白道》、《黑白战场》到《卧虎》，毫无悬念的，至电影的某个时段，必定要碰到"卧底"的情节，已经到了让人腻烦的地步。有影评家认为"卧底"系列是关于香港人"身份"问题的讨论，而我则认为不尽然。对于一个题材的反复使用，其实反映了在香港这个商业社会中的一种普遍社会心理：一个模式一旦成功，那么就要对其无限复制，以期达到相同的成功效应——典型的商业功利主义，从商业至电影、从设计至地产，无不奉为金科玉律，屡试不爽。对于成功模式的复制，后来者沉浸在"少投入，低风险，高收益"的窃喜中，却不知已经深陷入自己为自己设置的桎梏中——因为任何一种复制都是一次创意的贬值，复制的次数越多，它所引起的反感情绪就越大。

拷贝成功模式的做法在香港的中小学校舍设计中体现得最为明显。行遍全港，几乎所有的校舍都是一个模子里面套出来的。这与"标准学校"的政策难脱干系，这也是以上功利主义逻辑的体现。它甚至给人这样一种感觉：在香港，很多新建筑的建造可以不需要设计，完全按照某个建成的图纸复制即可。建筑师处于一种可有可无、普遍失语的状态。

而如果我们以荷兰的学校为参照，就可发现，荷兰的校舍规划逻辑完全建立在创造多样性个性化校园空间的宗旨上。例如乌德勒支大学校园设计中，每栋建筑尽可能在使用及形式上都有自我特色，而整体上仍然服从系统的要求。它的策略是整体规划由一个事务所（OMA）主持，而各个系楼则邀请不同的成名事务所进行个性化设计，这样就导向了一种可控制系统之下的多样性。所以乌大的新校园建设持续了十多年，每一件成品都堪称经典，却又在这么长的时间跨度中保持了连贯性。

较早期的经典为Herzberg设计的学生宿舍，可以看出结构主义的强烈影响，营造了一个低密度的社区。错动的体量、曲折的路径，虽然形式上仍然具有现代主义的简约风格，却完全摒弃了早期现代主义的冷漠与单调于近人尺度中，体现了一种人文关怀。

第二代经典落成于90年代，代表作为OMA的学生教育大厅。教育大厅是一个用单一表皮包裹的，各层之间的界限被模糊的平滑空间。可以适应多种用途：集会、演出、展览；只需添加少许装置，空间的性质就可以瞬间改变。它是一个无法被明确定义，却提供多种可能性的事件发生器。

而另一个个性鲜明的物理系楼着意在一个封闭的物体中创造静谧的感知空间。从底层的阴暗的入口楼梯进入之后，直达二层大厅，其左侧为一个占满整层的室内人造池塘，人工的暗泉涌动，在水面上泛起涟漪。它保持了内环境的湿度，并且借助微暗的光线，共同生发了一种平静的、可栖靠的场所：水声、倒影、微微湿润的空气。学生可以在"湖岸"或躺或坐，或三五成群的随意交谈。

最近落成的中心图书馆为威尔.阿雷兹设计，为一个黑色的方正容器，内部以各种开放式或半开放式的"阅览平台"飘浮于不同的高度，使得存书、阅读与交通空间可以时时渗透，打破了图书馆内部各种不同功能空间明确分割、相互隔绝的传统模式。密斯所提倡的"流动空间"在这里以三维的方式实现。

可见，校园建筑不应该是"标准化"的对象。"好用、简单、易管理"不应是校园建筑追逐的目标。处在人格形成重要阶段的青少年如果在千篇一律的僵化空间中学习、生活，如果他们的每日的环境就已经被标准化了，我们如何能期待他们日后可以做出更多的创意？因此即使在其他建筑类型领域，香港建筑师已经因绝望而放弃了创造，在校园建筑里仍然应当坚守一点起码的学科底线。

5. 消解过多的城市边界和非等级化

我们在"边界状态"的章节中讨论过，香港城市是一个"界面效应"无处不在的城市。从政府和地产商对于城市的控制，到各个利益团体对外界的设防和对于公众的排斥；从豪华住区周边的围墙樊篱等有形的边界直至对于空间领域化占有所造就的"无形边界"以及监视摄影机、会员制度等等多种"隐喻式边界"。大众已经越来越受到生活中的限制——没有突围的可能。而真正民主化的现代生活需要一个对于民众真正亲和和放松的都市环境，这是香港的建筑实践者所应解决的更深层次的都市问题——精神和心理层面的关注。

我们前面提及的荷兰海牙的Vinex是一个对于突破都市物理樊篱控制的有效尝试，在这个案例中，公共和私有空间之间没有明显的界限。通常的亚洲城市中，公共与私密的边界是以各种形式严格分开的：包括墙、樊篱、门禁制度、警卫、甚至狗（香港的住区是这种边界都市的典型之一）。而在海牙Vinex，公共空间与私有空间之间存在转化的可能。所以它的真正优势不在于住宅、花园和街道的统一化，而是公私之间物理边界的缺失。开放空间不再被划分为公共和私有，而是绿地与硬地、底层与高层形成三维的组合体。所有的空间都可自由进入，被各个公寓所共享。它导致了一种公众和私人对于"未定义"领域的灵活使用。我认为这正是海牙Vinex的关键突破所在。

另外在规划领域，现在欧洲有一种趋势是"少规划"理念的兴起。"少规划"是指规划师更多地注重城市自身在城市结果的形成中所起的作用，在城市规划的初级阶段尽量减少对于城市的过多的控制和过于明晰的功能指定，提供更多的自由度，让城市在自身发展中找寻其最合适的模式，并进行不断自我修正。

"少规划"的趋势在荷兰的逐渐增长，体现为一种自下而上的规划需求，允许多样性的功能和居住类型。在古典的荷兰城市中，街道存在等级：城市体块沿纬度方向布置；街区是封闭的，空出一两个街区作为小广场；私人花园与外部空间隔离。一战之后的街区，与中世纪相比，等级更加明显。宽阔的街道连接城市的各个角落，二级街道两边为商业，而三级街道则主要导向住宅。广场被模数化的两层联排住宅包围成不同的形状。在此之后，街区具有反组合的特征，街道丧失了作为公共空间的中性特征，广场成为开放空间，街区体块向三维方向发展，公共与私有之间的界限变得模糊。

而近年来的都市街区，则不再具有街道等级的特质。没有明确的道路，只有一条环路将区域与城市干道相连，汽车无法进入。根据规划者的说明，新建区域的居民多采用自行车。场地内为硬质铺地，车在外围直接驶入地下车库。在并置街区与公园的同时，形成了几栋建筑的孤岛。私人花园与底层公寓连接。这些变化也许可以成为亚洲都市借鉴的经验。

结 语：

香港城市是一个双重悖论的统一体：一方面，它拥有最有趣的城市，蕴涵着潜在的丰富性与多样性；而另一方面，它建筑设计方面极度平庸、商业化，并且缺乏创意。"不经意的自发城市结果的有趣"和"刻意设计的建筑的无趣"这一罕见的矛盾并存现象导致了这个城市在建筑学价值体系上同时兼具极好与极坏的属性而难以准确定义；欧洲的城市则正好相反，拥有具有相当热情和富有创意的设计师，也创造了完美的城市外表，却没有足够的城市活力，显示出一派"死气沉沉"的气象，因此城市的美好也成为一种虚伪的表象。

但是至少有一点可以肯定的就是：香港城市本身已经具有许多可以提供灵感的源泉，《隐形逻辑》对于这些都市的"特别之处"进行梳理，为设计提供可能的"新概念"的目录。香港的建筑，在创意方面，提升的空间仍然很大，而创意的终极目标仍然是人本主义。

参考书目

1. Rem Koolhaas. Delirious New York. New York: Monacelli Press, 1978
2. MVRDV. KM3, excursions on capacity. Barcelona:ACTAR , 2005
3. MVRDV. Far Max, excursions on density. Rotterdam: 010 Publishers , 1998
4. Rem Koolhaas. Mutations, Harvard project on the city. Barcelona: ACTAR, 2000
5. Le Corbusier. Toward a New Architecture. New York: Dover Publications, 1986
6. Gilles Deleuze. A Thousand Plateaus, Capitalism and Schizophrenia:University of Minnesota   Press, 1981
7. Martin Heidegger. The Question Concerning Technology. New York: Haper & Row, 1977
8. 陈翠儿,陈建国,蔡宏兴等. The 逼 City. 香港特区民政事务局，2006
9. 胡恩威. 香港风格. TOM Publishing Ltd. 2005
10. 古儒郎,林海华. 香港词典. 城市中国, 2005(12)

后 记

在向贝尔拉格学院的一位同仁讨论研究的主要内容时,被问及一个问题:是否研究中所描绘的现状就是最好的状态?城市是否可以变得更好?这个问题实际上是在追问"这个研究的批判性如何体现?作者的立场是什么?"我想在全文之末对研究的目的做一个回溯式诠释。

1. 关于"立场"问题使我联想到不久前,建筑界一场热闹的关于"批判与后批判"之争以及其引发的相关讨论。在我看来,非左即右的极端判断已经不合时宜——或者说这两种传统标准已经无法准确定义今天我们所面对的城市状况。因为现代主义者坚持的作为"以社会关怀为原始目的"的现代主义在晚期显示出力不从心,它恰恰失败于其"非人性"——其原初所设定的目标——即使"其内核是正当的";而另一方面,商业建筑师对于资本的无情拥抱和妥协,使我们的城市充满了大量商业化的平庸的建筑垃圾,这两种态度都无法解决今天的问题——我们需要第三种道路。

建筑师的首要任务是如何对于现实的问题提出创造性的解决方式,如何以建筑挑战既有的成规、教条与不完善的生活积习,从而产生主导新的生活方式,或者改变固有意识形态的可能。设计的本质是一种创造性思考,而当今建筑师只能通过奇形怪状的形式讨论来寻找灵感。如果建筑学的内涵仅仅止于这个层面,它基本上已经失却了发展的动力。

《隐形逻辑》虽然基于现实,但是它不仅仅是一种"再现现实"。它提倡一种在平庸中发现"特别"的洞察力,它从现实中找寻概念,为设计的创新积累一种"理论势能"。建筑不是简单的关于"对错",而是关于"是否有趣"。

2. 另一点考量是基于对中国建筑界的现状的隐忧——建筑师们埋首于设计,已经直接达到了"都市人工物"的操作层面。可是,在动手之前,却忽略了对于所要操作的对象——城市的"了解"。也许是过速的城市化使我们无暇了解,也许是研究本身费时却与经济效益无太大关联使我们不屑于了解。但是,不了解则可能导致盲目,不了解城市即开始设计、建造,就如同医生尚未查明病情就开始给病人动手术。所以今天的城市结果难免满目疮痍,难以让大众满意。

笔者希望这本小书可以作为一个起始之"砖",引发出一种研究的氛围,为建筑师缺失的身份定义进行完整化重塑——即建筑师除了设计之外,至少还应该是一个城市研究者。中国如火如荼的城市化带来的启示与谬误一直并存滋长。在西方学界纷纷把目光投向我们的城市时,本土建筑师不应该仅仅做一个旁观者,因为城市本身已经为我们提供了大量可供推敲与反思的现实素材。

3. 我们所常见的城市文本,多数的叙述方式采用的是"文字+照片"的形式,其中缺乏了一个重要环节——"分析"。作为建筑师,图解是一种表达的有效手段。本研究尚且希望在城市论述的表达方式方面,做一些探索和尝试。

帷幕已经拉开,而正剧才刚刚开始……

作 者

2008年12月于香港

作者简介

张为平

荷兰注册建筑师；
荷兰代尔夫特理工大学建筑学硕士；
曾于荷兰的MVRDV及比利时的BURO2事务所工作；
2008于香港从事建筑实践以及高密度状态下的城市研究；
2009年成立城市研究工作室"都市可能概念工厂"
坚信建筑应当比现状中的能量更大，
并试图为其找寻一个释放的出处……

Zhang Weiping
Dutch Registered Architect;
Master of science in Architecture, Delft University of Technology;
With experience in MVRDV Rotterdam, BURO2 Brussels,
Presently practicing in Hong Kong, simultaneously researching the city in high density condition;
co-founder of the studio "Idea Factory of Urban Possibilities"(IFUP)
Believing that the architecture should be more powerful than what it reveals now,
And trying to find a way to explore its hidden potential.

对于本研究有任何建议请联系我们：
ideafactory_up@163.com